Plant Biotechnology and Medi

Mohamed Ramadan Rady

Plant Biotechnology and Medicinal Plants

Periwinkle, Milk Thistle and Foxglove

Springer

Mohamed Ramadan Rady
Department of Plant Biotechnology
National Research Centre
Giza, Egypt

ISBN 978-3-030-22928-3 ISBN 978-3-030-22929-0 (eBook)
https://doi.org/10.1007/978-3-030-22929-0

This Springer imprint is published by the registered company Springer Nature Switzerland AG
The registered company address is: Gewerbestrasse 11, 6330 Cham, Switzerland

Dedicated to my beloved family, Ayah, Achraf, and Salwa

Preface

Recently, secondary metabolite production through plant cell culture and plant biotechnology has attracted interest from scientists. The potential of some plant culture systems for the production of medically important compounds has been demonstrated. Recently, genetic transformation and metabolic engineering are important areas which may provide new ways and efficient systems to increase in vitro production of secondary metabolites in medicinal plants. This system can now provide a commercially realistic alternative to whole plants for the production of some drugs. More recently, the progress on the genetic manipulation of biosynthetic units in microorganisms (synthetic biology) has opened the possibility of in-depth exploration of the large chemical space of natural product derivatives. In spite of several successful reports on the studied plants, there is still a gap in the knowledge for biosynthetic pathways and upscaling of this culture system for commercial utilization.

The purpose of this book is to provide recent information and studies about the induction of cell and organ cultures, establishment of transgenic cultures, and induction of hairy roots from in vitro cultures of periwinkle, milk thistle, and foxglove plants with special emphasis on elicitation strategies by abiotic and biotic elicitors for the production of the bioactive compounds from cultures. This book is expected to serve as a guide for the use of plant biotechnology as alternative source for increasing pharmaceutically important anticancer, flavonolignan, and cardenolide compounds from the studied plants. This book will be valuable to researchers as well as students working in the area of medicinal plant biotechnology. It will also serve as a reference for the pharmaceutical industry.

Giza, Egypt Mohamed Ramadan Rady

Acknowledgment

Foremost, I would like to thank Allah, without whom, nothing is possible. I dedicate this work to my deceased parents who gave me the best gift of all, higher education and the freedom of choice. Also, I owe a special thanks to my brothers for their love and unwavering support throughout my life and to Dr. Kenneth Teng, the editorial staff of Springer, New York, for his guidance and encouragement and for giving me the opportunity to conduct this book, without him, there would be no book at all. I would also like to acknowledge the Springer team for their help in technical and continuous support and for editing this book. Last, but not least, I would like to gratefully acknowledge the National Research Centre, Egypt, for their constant support and excellent care.

Mohamed Ramadan Rady

Contents

Abbreviations

ABT	1-aminobenzotriazole
ACES	2-[(2-amino-2-oxoethyl)amino]ethanesulfonic acid
$AgNO_3$	Silver nitrate
APX1	Ascorbate peroxidase 1
AS	Anthranilate synthase
AVG	Aminoethoxyvinylglycine
B5	Gamborg et al. (1968) medium
BA	Benzyl adenine
CA	Coniferyl alcohol
$CaCl_2$	Calcium chloride
CAD	Cinnamyl alcohol dehydrogenase
$Cd(NO_3)_2$	Cadmium nitrate
CDPK	Calcium-dependent protein kinases
CDs	Cyclodextrins
CH	Casein hydrolase
CHI	Chalcone isomerase
CHS	Chalcone synthase
chsA	Chalcone synthase gene
CLOT	Clotrimazole
CPA	Chlorophenoxyacetic acid
$CuSO_4$	Copper sulfate
D4H	Deacetoxyvindoline 4-hydroxylase
DMAPP	Dimethylallyl diphosphate (DMAPP)
DMSO	Dimethyl sulfoxide
EGTA	Ethylene glycol tetraacetic acid
ER	Endoplasmic reticulum
$FeSO_4$	Iron sulfate
G(G)PPS	Geranyl (geranyl) diphosphate synthase
G10H	Geraniol 10-hydroxylase
GA_3	Gibberellic acid
GES	Geraniol synthase

GPP	Monoterpene geranyl diphosphate
H_2O_2	Hydrogen peroxide
H_3B0_3	Boric acid
$HgCl_2$	Mercuric chloride
HPLC	High-performance liquid chromatography
IAA	Indole acetic acid
IBA	Indole butyric acid
IPP	Isopentenyl diphosphate (IPP)
JAs	Jasmonates
KCl	Potassium chloride
KI	Potassium iodide
$LaCl_3$	Lanthanum
LiCl	Lithium chloride
LN	Liquid nitrogen
LS	Linsmaier and Skoog (1965) medium
MBPK	Myelin basic protein kinase
MDA	Malondialdehyde
MeJa	Methyl jasmonate
MEP	Methylerythritole phosphate
MES	2-(N-morpholino) ethanesulfonic acid
$MgSO_4$	Magnesium sulfate
mM	Millimolar
$MnSO_4$	Manganese sulfate
MS	Murashige and Skoog (1962) medium
Mst	Mastoparan
Na_2MoO_4	Sodium molybdate
NAA	Naphthalene acetic acid
NaCl	Sodium chloride
nBuOH	n-butanol
Ng	Flavanone naringenin
NmI	Lehmann et al. (1995) medium
NN	Nitsch and Nitsch (1969) medium
NO	Nitric oxide
NTPII	Neomycin phosphotransferase II gene
OD	Optical density
P450scce	Cholesterol side-chain cleavage enzyme
P5βR	4 progesterone-5β-reductase
PAL	Phenylalanine ammonia lyase
PCR	Polymerase chain reaction
PCA	2,5-pyridinedicarboxylic acid
PEG	Polyethylene glycol
PLB	Protocorm-like body
PLD	Phospholipase D
POD	Peroxidase
ppm	Parts per million

PUF	Polyurethane foam
ROS	Reactive oxygen species
SAAT	Sonication-assisted Agrobacterium-mediated transformation
SAR	Systemic acquired resistance
SCCE	Sterol side chain cleaving enzyme
SGD	Strictosidine-β-dglucosidase
SH	Schenk and Hildebrandt (1972) medium
SNP	Sodium nitroprusside
SOD	Superoxide dismutase
$SrCl_2$	Strontium
STB	Stirred-tank bioreactors
STR	Strictosidine synthase
STS	Stilbene synthase
TAB	Tetramethylammonium bromide
TDC	Tryptophan decarboxylase
TDZ	Thidiazuron
TIA	Terpenoid indole alkaloid
TRIA	Triacontanol
TTC	Triphenyltetrazolium chloride
Tx	Taxifolin
UV	Ultraviolet
$VaSO_4$	Vanadyl sulfate
WPM	Loyd and McCowan (1980) medium
YE	Yeast extract
$ZnSO_4$	Zinc sulfate
μM	Micromolar
μg	Microgram
2,4-D	2,4-Dichlorophenoxyacetic acid
3-KSI	3 Δ4,5-3-ketosteroid-isomerase
3β-HSD	3β-hydroxysteroid dehydrogenase

About the Author

Mohamed Ramadan Rady is a Professor Emeritus at Plant Biotechnology Department, Genetic Engineering and Biotechnology Division, National Research Centre, Egypt. He has obtained his doctoral degree in plant biotechnology from Agronomy Department, College of Agriculture, Al-Azhar University, Cairo, Egypt. His basic area of research is plant biotechnology (plant secondary metabolites and genetic manipulation of medicinal plants), and he did work on several medicinal and aromatic plants. To date, he has successfully run bilateral research project as Principal Investigator on periwinkle plant with The Hungarian Academy of Sciences. He has also participated in various national and international conferences. He is an Associate Editor of *Journal of Genetic Engineering and Biotechnology* (Springer Nature), and reviewer of many national and international scientific journals. He is involved in research, teaching, training, and supervising research students at the department and other universities. Based on his research contributions, he has been awarded a prize of the National Research Centre for scientific encouragement in the field of agriculture and biological sciences. Dr. Rady has published more than 45 research articles on medicinal plants in reputed journals. He has authored one book chapter and edited this book.

Chapter 1
Plant Biotechnology and Periwinkle

Contents

M. R. Rady, *Plant Biotechnology and Medicinal Plants*,
https://doi.org/10.1007/978-3-030-22929-0_1

Abstract *Catharanthus roseus* is one of the most extensively investigated medicinal plants, which can produce more than 130 alkaloids, including the powerful antitumor drugs vinblastine and vincristine which are used in the treatment of cancer. Alkaloids are one of the most important secondary metabolites known to play a vital role in various pharmaceutical applications leading to an increased commercial importance in recent years. An overview of recent studies which have been used using various approaches of plant tissue, organ culture, regeneration, cryopreservation, and transformation of *C. roseus* is presented in this chapter. One of the most effective strategies for enhancing the biotechnological production of alkaloid compounds is elicitation. This chapter summarizes the recent research work of various in vitro cultures, abiotic, biotic elicitors, and precursor feeding applied to *C. roseus* cultural system and their stimulating effects on the accumulation of TIAs.

Keywords *Catharanthus roseus* · Biosynthesis · Cell and callus culture · Regeneration · Cryopreservation · Transformation · Hairy root cultures · Elicitation · Alkaloids

1 Introduction

Catharanthus roseus is an important medicinal plant, belonging to the family Apocynaceae, and is a rich source of alkaloids, which are distributed in all parts of the plant. *C. roseus* is a perennial, evergreen herb that was originally native to the island of Madagascar. It has been widely cultivated for hundreds of year and has been widely cultivated in all warm and pantropical regions of the world (Aslam et al. 2010). The subshrub grows about 30–100 cm high with glossy and dark green leaves of 2–5 cm long and 1–3 cm broad. The wild *C. roseus* plant has a pale pink phlox-like flower with a purple eye in the center, but various cultivars have been developed with flower colors ranging from purple, violet, red, pink, and white (Fig. 1.1).

Catharanthus roseus contain important alkaloids, viz., vincristine and vinblastine, which are used in the treatment of cancer. These alkaloids interfere with the mitotic cell division process of the cancerous cells. They stop formation of microtubules, and thus chromosomes are unable to arrange on metaphase plate (Negi 2011). The vinblastine and vincristine can lower the number of white cells in blood. A high number of white cells in the blood indicate leukemia. So they act as anticancer drug. These alkaloids prevent mitosis in metaphase, and they bind to tubulin and thus prevent the cell from making the spindles it needs to divide (Kalidass et al. 2010). However, ajmalicine, serpentine, vindoline, and catharanthine are major alkaloids. In these alkaloids, ajmalicine and serpentine are useful for treatment of hypertension. Vindoline and catharanthine are the obvious precursors in the biosynthetic pathways of dimeric indole alkaloids such as vinblastine and vincristine. But their isolation from intact *C. roseus* plants is very costly because of their extremely low concentrations. Alkaloids from the *C. roseus* are

Fig. 1.1 *C. roseus* plant

normally obtained from the field-grown plants. It requires lots of space and infrastructure; in addition the raw material is season dependent and is affected by various fluctuating environmental risk factors. The antitumor alkaloids are produced in trace amounts (0.0003% dry weight) (Negi 2011). The high prices of these anticancer products, ranging from $1 million to $3.5 million per kilogram, have led to a widespread research interest over the past 25 years in the development of alternative sources for the production of these compounds (Verpoorte et al. 1991). The low yield and high market price of the pharmaceutically important alkaloids of *C. roseus* have created interest in improved alternative routes for their production (Verma et al. 2012). However, most TIAs are present in very small amounts, especially the dimeric/bisindole alkaloids. Thus large quantities of raw material are needed for compound isolation. For example, to isolate 1 g of vinblastine, about 500 kg of *C. roseus* leaves are required (Van der Heijden et al. 2004). In addition, it is also difficult to synthesize TIAs by chemical methods due to their complicated structures (Yang and Stöckigt 2010). However, the yields of these TIAs are low in wild-type plants, and the total chemical synthesis is impractical in large scale due to high cost and their complicated structures (Wang et al. 2012).

As the demand for medicinal plants is growing at a very fast pace, consequently some of them are increasingly being threatened even in their natural habitats (Muthukumar et al. 2004). For these reasons, a biotechnological approach using plant cell or tissue cultures is being explored as alternative production method of the valuable bioactive metabolites from plants. Plant tissue culture might be a source of these monomeric and dimeric alkaloids, and therefore, many attempts have been made to establish a culture that produces them in large amounts. Ajmalicine, serpentine, and catharanthine are produced by some cell culture lines in amounts several times those obtained from intact plant (Kurz et al. 1981). Researchers are focusing their attention to enhance the alkaloids yield by various ways, chemically, enzymatically, and synthetically, or by cell culture method.

In this chapter, recent studies for establishment of cell, callus, organ, and transgenic cultures were reviewed. Optimization of the culture medium, plant growth regulators, and culture conditions were extensively studied to improve the cell biomass accumulation and the TIA production. However, abiotic, biotic elicitors and precursor feeding to enhance accumulation of TIAs in different cultures of *C. roseus* are highlighted.

2 Biosynthesis of TIA in *C. roseus*

Alkaloids of *C. roseus* comprise a group of more than 130 terpenoid indole alkaloids (TIAs) which represent one of the largest and most diverse groups of alkaloids in this plant. It is the sole resource of vinblastine and vincristine, which are two of the biggest concerns of TIAs because of their powerful anticancer activities (Van Der Heijden et al. 2004). A complete knowledge of the biosynthetic pathway of the targeted compounds and its regulation is essential to increase the metabolic flux toward the desired products. The biosynthesis of *Catharanthus*' alkaloids has been studied extensively. The biosynthetic pathway of TIAs in *C. roseus* and characterization of the related genes encoding the enzymes involved in this pathway are summarized in Fig. 1.2.

The biosynthesis of TIA in *C. roseus* is a complex metabolic pathway involving different subcellular compartments including plastids, cytosol, nucleus, endoplasmic reticulum (ER), and vacuole, in which biosynthetic machinery lies within membranes for alkaloid metabolism (Pomahacova et al. 2009). More than 50 biosynthetic events are composed of the involved genes, enzymes, regulatory genes, and intra-/intercellular compartments (Zhao et al. 2013). TIA biosynthesis requires two precursors from two different biosynthetic routes, i.e., tryptamine from the shikimate/tryptophan pathway and secologanin from the iridoid/methylerythritol phosphate (MEP) pathway (Pan et al. 2016).

Among the two precursor pathways, the iridoid pathway is considered a major rate-limiting factor for TIA production in *C. roseus* cell cultures (Zhao and Verpoorte 2007; Pan et al. 2016). The iridoid precursors of the TIA derive from 8-hydroxygeraniol (also known as 10-hydroxygeraniol) which is formed upon hydroxylation of geraniol generated from monoterpene geranyl diphosphate (GPP). GPP is a condensation product of the basic isoprene units, isopentenyl diphosphate (IPP) and dimethylallyl diphosphate (DMAPP). The other enzymes such as geraniol 10-hydroxylase (G10H), NADPH-cytochrome P-450 reductase, and anthranilate synthetase (AS) have the similar TDC activities which are involved in the biosynthesis of indole alkaloids.

Formation of Tryptamine and Secologanin

Tryptamine is derived from indole biosynthetic pathway. Tryptamine is derived from a single enzymatic conversion of the amino acid L-tryptophan (product of

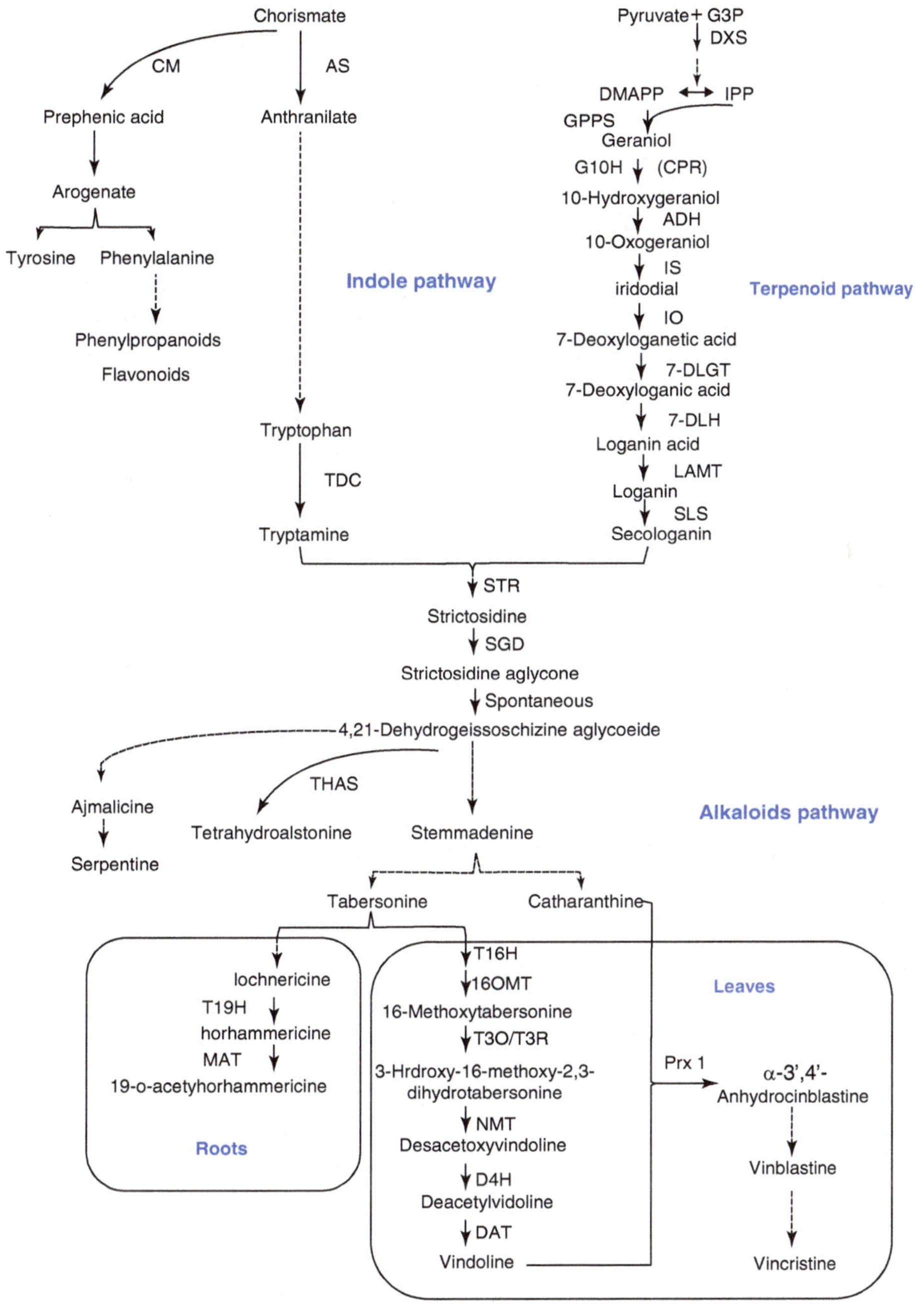

Fig. 1.2 Terpenoid indole alkaloid pathway. (Dashed lines represent the unknown steps) (Adapted from Sun et al. 2016)

the plastidial shikimate pathway) by the enzyme tryptophan decarboxylase (TDC) (De Luca et al. 1989), while secologanin is derived from the monoterpene geranyl diphosphate (GPP) in the plastidial methylerythritol phosphate (MEP) pathway (Contin et al. 1998b; Hong et al. 2003). Strictosidine synthetase (STR) catalyzes the coupling of tryptamine and secologanin to produce strictosidine. These two compounds are the universal precursors of all TIAs in plants.

Biosynthesis of Strictosidine

Strictosidine is the central intermediate in the biosynthesis of many alkaloids in *C. roseus*, which is derived from the condensation of secologanin and tryptamine by strictosidine synthase (STR) (Fig. 1.2). Strictosidine synthase (STR) was shown to be localized in the vacuole; thus tryptamine and secologanin from the MEP pathway need to be transported to the vacuole to produce strictosidine. Subsequently, strictosidine is transported out of the vacuole to be deglucosylated by strictosidine-*β*-dglucosidase (SGD) which is associated with the nucleus (Guirimand et al. 2010). The reversible intermediate 4,21-dehydrogeissoschizine can also be converted into stemmadenine which leads to the production of vindoline and catharanthine, the monomeric precursors of bisindole alkaloids vinblastine and vincristine (El-Sayed and Verpoorte 2007).

Biosynthesis of Vinblastine and Vincristine

Biosynthesis of vinblastine and vincristine involves a series of enzymatic reaction localized in the endoplasmic reticulum (ER) (tabersonine 16-hydoxylase [T16]), cytosol (16-hydroxytabersonine 16-*O*-methyltransferase [OMT], desacetoxyvindoline 4-hydroxylase [D4H], and deacetylvindoline 4-*O*-acetyltransferase [DAT]), thylakoid membrane of chloroplasts (*N*-methyltransferase [NMT]), and vacuole (peroxidase [PRX1]) (Costa et al. 2008; Guirimand et al. 2011). The bisindole alkaloids vinblastine and vincristine are derived from the coupling of the monomeric alkaloids catharanthine and vindoline; the process is catalyzed by the major class of vacuolar III peroxidase (CrPrx1) (Costa et al. 2008). These monomeric alkaloids produced anhydrovinblastine by a peroxidase which is a reduction product. The product α-3′,4′-anhydrovinblastine catalyzed by CrPrx1 is the common precursor of all dimeric alkaloids, which can be future converted into vinblastine and vincristine through several steps (Zhu et al. 2015).

In general, the TIAs are condensation products of two biosynthetic routes, requiring coordination of the amount of intermediates supplied by both pathways. Starting from the amino acid tryptophan and the monoterpenoid geraniol, the biosynthesis of vinblastine requires the participation of at least 35 intermediates, 30 enzymes, 30 biosynthetic, 2 regulatory genes, and 7 intra- and intercellular compartments. Although many genes and corresponding enzymes have been characterized in this pathway, our knowledge on the whole TIA biosynthetic pathway still remains largely unknown up to date. Full elucidation of TIA biosynthetic pathway is an important prerequisite to understand the regulation of the TIA biosynthesis in the medicinal plant and to produce valuable TIAs by synthetic biological technology (Zhu et al. 2014).

3 In Vitro Culture of *C. roseus*

3.1 Cell and Callus Cultures of C. roseus

Various explants (stem, leaf, hypocotyl shoot tips, roots, and petioles) were used from in vitro germinated seedlings or from in vivo plants of *C. roseus* which have been tested as primary explant sources on several basal media with plant growth regulators for obtaining callus cultures. It has been noted that the plant cell's growth is fast in agitated suspension compared to solid medium because of easier uptake of nutrients by the cells. Establishment of callus and cell suspension cultures in *C. roseus* is considered the first step to study biosynthetic capacity and improve the production of TIAs. The procedures of standardizing friable calli or suspension culture development become a necessary step to manufacture valuable plant metabolites.

Several studies have investigated the effects of plant growth regulators, culture media, and cultural conditions for establishment of *C. roseus* cultures. In vitro studies of cell and callus culture induction from *C. roseus* were presented in Table 1.1.

Miura and Hirata (1987) obtained callus cultures of greenhouse-grown *C. roseus* plants. They found that MS medium supplemented with 1.0 mg/l NAA and 0.1 mg/l kinetin induced callus tissues from young leaf explants. Initiated calli were brown and tight and differentiated many hairy white roots and maintained high antimitotic activity for several passages. However, isolation of vinblastine in callus culture with differentiated roots was detected. Zhao et al. (2001a) observed that there was no obvious difference between the compact callus cluster cultures derived from leaf explants and stem explants from *C. roseus* grown on MS liquid induction medium supplemented with 5.37 μM NAA and 4.65 μM kinetin. They also postulated that the level of cellular/tissue differentiation might be responsible for these different alkaloid synthesis capabilities. Sucrose regime affected some properties (the size, degree of compaction, differentiation level) of the compact callus cluster cultures and therefore influenced alkaloid production. The optimal sucrose concentration for alkaloid and biomass production by that cultures was 50 or 60 g/l. Junaid et al. (2006) found that MS medium supplemented with 1.0 mg/l 2,4-D was the best for callus induction from hypocotyl explants of *C. roseus*. The maximum callusing response percentage was 85. The hypocotyl callus was friable, light yellow, and fast growing. In another study, callus cultures were initiated from single-node explants of field-grown *Vinca minor* when cultured on MS medium supplemented with the combination of 7.21 mg/l BAP and 2 mg/l NAA (Raouf Fard et al. 2008). Also, callus cultures were initiated from petiole explants of *C. roseus* greenhouse-grown plants, when cultured on MS medium containing 0.1 mg/l NAA + 0.1 mg/l Kin after 6-week incubation (Ataei-Azimi et al. 2008).

Four aseptically explants (shoot tip, leaf, stem, and root) were excised from the plantlets as well as sterilized seeds of *C. roseus* and used for callus initiation (Taha et al. 2008). They found that response of shoot tip explant was the best between the different explants on MS medium supplemented with 1.0 mg/l 2,4-D and 1.0 mg/l

Table 1.1 Studies on cell and callus cultures induction of *C. roseus*

Explant used	Medium used	Culture condition	Growth regulators (mg/l)					Reference
			2,4-D	NAA	Kin	BA	TRIA	
Young leaf segments of greenhouse-grown plants	MS	Cultures were incubated at 25 °C in the dark		1.0	0.1			Miura and Hirata (1987)
Stem and leaf explants from cultivated plants in a greenhouse	MS	Cultures were incubated on a rotary shaker (120 rpm) at 23–28 °C in darkness		5.37 μM	4.65 μM			Zhao et al. (2001a)
Hypocotyl of 5–7 days old in vitro germinated seedlings	MS	Cultures were incubated under a 16 h photoperiod at 25 °C light/20 °C dark	1.0					Junaid et al. (2006)
Single node	MS	All cultures were incubated at 25 ± 2 °C under a 16 h photoperiod		2		7.21		Raouf Fard et al. (2008)
Petiole segments from leaves of 6-week-old seedlings, germinated in greenhouse	MS	Cultures were placed in a growth chamber at dark period for first of 2 weeks and 24 h light period for second of 4 weeks at 35 °C		0.1	0.1			Ataei-Azimi et al. (2008)
Shoot tip from in vitro germinated seedlings	MS	Cultures were incubated in a growth chamber at 25 ± 2 °C under a 16/8 h light dark photoperiod	1.0		1.0			Taha et al. (2008)
Shoot tips collected from the field-grown plants	MS	Cultures were maintained in dark for 4–6 weeks at 25 ± 30 °C	2.0 μM				5.0 μM	Malabadi et al. (2009)
Hypocotyl, leaf, nodal, and root from in vitro germinated seeds	MS	Cultures were incubated at 25–28 °C under 16 h photoperiod	6.96 μM					Aslam et al. (2009)
Leaves were harvested from plants growing in gardens	NN	Different light conditions		8.0 μM		8.0 μM		Al-Khateeb et al. (2009)
Stem explant from mature plants	MS	Cultures were maintained at 25 ± 1 °C with a 16 h photoperiod	1.0 μM		1.0 μM			Kalidass et al. (2010)
Hypocotyl explants from 10–15-day-old seedlings grown in greenhouse	MS	Cultures were maintained at 26 °C with completely dark conditions		1.0		1.0		Singh et al. (2011)

Leaf explants	MS	Dark conditions		2.0		3.0		
Epicotyl explants	MS	Dark conditions	2.0			0.1		
Root explants	MS	Dark conditions		1.0		1.5		
Segment of shoots and roots from in vitro germinated seeds	B5		1.0		1.0			Fathalla et al. (2011)
	MS Subculture				1.0			
Leaves were collected from third node from the apex of garden-grown plants of *Vinca rosea*	MS	Cultures were incubated in a light regime of 18 h followed by 6 h dark period at 25 ± 2 °C	1.0		0.1			Negi (2011)
Nodal segment were collected from garden-grown plants of *Vinca rosea*	MS				1.0	1.0		
Young leaves, from mature plant	MS	Cultures were maintained in a culture room at 18 h photoperiod at 25 °C		1.25		0.2		Zulkepli and Samad (2011)
Young shoot tip from plants cultivated in a greenhouse	MS	Cultures were incubated at 25 ± 1 °C under complete darkness	1.5		0.5			Saifullah and Khan (2011)
		Cell suspension cultures on a gyratory platform shaker at 100 rpm and 25 ± 1 °C with a 16 h photoperiod						
Leaf explants from in vitro germinated seedlings	MS	½ MS; cultures were incubated under a 16 h photoperiod at 25 ± 20 °C	0.5			1.0		Verma et al. (2012)
Leaves that are still actively growing at 3–4 leaves from the apex shoots were used as the explants	MS	Cell aggregate cultures were incubated at room temperature and agitated at a speed of 120 rpm		2.0	0.2			Pandiangan et al. (2013)

(continued)

Table 1.1 (continued)

Explant used	Medium used	Culture condition	Growth regulators (mg/l)					Reference
			2,4-D	NAA	Kin	BA	TRIA	
Hypocotyls of in vitro germinated seeds	MS	Cultures were incubated under a 16 h photoperiod at 25 °C light/20 °C dark	1.0					Aslam et al. (2014)
Various parts (stem, leaf, and hypocotyl) were used from germinated seedlings	MS	Cultures were incubated under a 16 h photoperiod at 25 °C light/20 °C dark	4.52 μM					Mujib et al. (2014)
Leaf explants from in vivo plants	MS + 1.0 gm/l casein hydrolysate + 0.6% agar			1.0		1.0		Veerabathini et al. (2015)
Leaves and stem of plants from greenhouse-grown plants	MS	Cultures were incubated under darkness in a culture chamber at 25 °C for callus induction	2.5					Rashmi and Trivedi (2015)
Leaves from plants grown in greenhouse	MS	Cultures were placed in a growth chamber at 25 ± 2 °C and photoperiod 16/8 h light/dark	1.0			1.0		Khashan and Husain (2015)
Leaf explants of cultivated plants in a greenhouse	Vermicompost extract only							Kashyab and Kale (2015)
Nodal segment from mature plants	MS	Cultures were incubated at 16:8 h photoperiod and temperature of 27 ± 20 °C		2.0		1.0		Moghe et al. (2016)
Leaves of plants from greenhouse-grown plants	MS	Cultures incubated at 25 ± 2 °C and photoperiod 16/8 h light/dark	1.0			1.0		Khashan and Al-Athary (2016)
Hypocotyl explants from in vitro germinated seedlings	MS	Cultures were maintained at 25 ± 2 °C under a 16 h light and 8 h dark regimes		0.5		1.0		Alam et al. (2017)

Cotyledons excised from in vitro grown seedlings (cv. Nirvana Pink Blush)	MS	Cultures incubated in darkness at 25 ± 2 °C		1.0		2.0		Al-Zuhairi and Ghanm (2017)
Leaves of the third or fourth of the shoots	MS		2.0		0.2			Mandagi et al. (2017)
Shoot tips from in vivo plants with white flower	Liquid MS	Incubated in the dark at 25 ± 2 °C on orbital shaker at 100 rpm	1.5					Pliankong et al. (2018)
	Liquid MS	Incubated under 16 h photoperiod at 25 ± 2 °C on orbital shaker at 100 rpm	1.5		0.5			
Leaves excised from in vitro grown seedlings	MS	Cultures were kept at 23 ± 1 °C in dark condition	1.5			1.5		Mekky et al. (2018)

MS Murashige and Skoog medium (1962), *B5* Gamborg and Shyluk (1981), *NN* Nitsch and Nitsch (1969), *TRIA* triacontanol

Kin. It was found that callus fresh weights of different explants were gradually increased by increasing the time of cultivation. Malabadi et al. (2009) induced friable embryogenic callus from thin sections of shoot tips collected from the field-grown plants of *C. roseus* and grown on MS medium supplemented with 2.0 μM 2,4-D and 5.0 μM triacontanol. Also, Aslam et al. (2009) reported that, for induction of callus cultures, the explants (leaf, nodal segments, and root) from *C. roseus* were cultivated on MS supplemented with 6.96 μM 2,4-D. Al-Khateeb et al. (2009) found that NN medium containing 8.0 μM each of BAP and NAA was the best for callus formation from leaf explants of *C. roseus* growing in gardens.

Leaf and stem segments from mature plants of *C. roseus* were cultured on MS medium with different growth regulators (Kalidass et al. 2010). The results showed that the overall response to plant regulators in stem segments was superior, while leaf explants resulted in poor callus induction. Friable greenish-yellow or beige callus was successfully induced from wound sites in the young stem explants at a culture time in the range of 5–10 days. MS medium supplemented with 2,4-D (1.0 μM) and kinetin (1.0 μM) was used to support the growth of callus cultures, and the maximum amount of dry biomass (598.04 mg) was produced after 7 weeks of culture.

Singh et al. (2011) showed that the highest percentage (92.2 and 94.9) of callus induction was obtained from hypocotyl explant of *C. roseus* grown on MS medium supplemented with 1.0 mg/l BAP + 1.0 mg/l NAA under light and dark conditions, respectively. Maximum callus induction from epicotyl explants was observed on medium supplemented with BAP (0.1 mg/l) + 2,4-D (2.0 mg/l) with percentage ranged between 64.1 and 90.8 under light and dark conditions respectively. For leaf explants, maximum callus induction response (90.2% and 92.2%) was obtained on MS-medium supplemented with BAP (3.0 mg/l) + and NAA (2.0 mg/l) under light and dark conditions, respectively. MS-medium supplemented with BAP (1.5 mg/l) + NAA (1.0 mg/l) was the best for callus induction percentage (71.4 and 74.6) from root explants under both light and dark conditions, respectively. Fathalla et al. (2011) reported that B5 (Gamborg and Shyluk 1981) medium supplemented with 1.0 mg/l each of 2,4-D and Kin was the best for callus initiation from segment of shoots and roots of *C. roseus.* Also, Negi (2011) found that MS medium supplemented with kinetin and BAP each of 1.0 mg/l and 2,4-D and IAA each of 1.0 mg/l combinations showed good callus production from nodal explants of *C. roseus*. When the leaf explants were cultured in MS + kinetin (0.1 mg/l), 2,4-D (1.0 mg/l), MS + BAP (0.1 mg/l), and IAA (1 mg/l), a good callus tissue was induced.

Zulkepli and Samad (2011) found that the highest frequency of callus formation was obtained from *C. roseus* leaf explants (1.3 g) (100%) at a combination of 0.2 mg/l BA with 1.25 mg/l NAA. In another study, callus was produced from young shoot tip on MS medium supplemented with 1.5 mg/l 2,4-D and 0.5 mg/l Kin and solidified by 8 g/l agar. Off-white callus with friable texture was found to be the best when this callus transferred into the liquid medium; it starts disintegrating easily making good suspension culture. Suspension culture was produced from induced callus on the same medium except Kin and agar (Saifullah and Khan 2011).

Verma et al. (2012) reported that half-strength MS basal medium supplemented with 2,4-D (0.5 mg/l), BA (1.0 mg/l), and 6% sucrose was best for biomass production of leaf callus and enhancement of alkaloid accumulation in *C. roseus*. Pandiangan et al. (2013) induced callus cultures from leaf explants of white flower *Vinca*, *C. roseus*. They initiated cell aggregate cultures when callus tissues were transferred to liquid MS medium with 2.0 mg/l NAA and 0.2 mg/kinetin. Cell aggregates yielded optimum cell growth on days 14–21 with tryptophan treatments 100 and 150 mg/l. The highest wet weight (3.63 g) was also achieved with tryptophan (150 mg/l) that was equal to 77.75% (increase the percentage of wet weight) and occurred on day 14. Aslam et al. (2014) used hypocotyls of in vitro germinated seeds of *C. roseus* as explant for initiation of callus tissues. MS medium supplemented with 2,4-D (1.0 mg/l) induced white to yellowish callus within 8–10 days of incubation. The callus growth was prolific, which later turned into embryogenic callus within 4 weeks of culture.

For embryogenic callus initiation, Mujib et al. (2014) reported that all the plant parts from *C. roseus* responded well in culture and produced callus on auxin-supplemented solid MS medium, but the MS medium supplemented with 4.52 μM 2,4-D was the best for hypocotyl explant. The calli induced from other plant parts were non-embryogenic which were compact, hard nodular structure and grew slowly. The hypocotyl callus was friable, light yellow, and proliferated well on solid medium. Higher levels of auxin inhibited callus induction and subsequent callus growth. They also found that the callus biomass growth was however relatively low in solid compared to liquid MS medium. The callus biomass in solid was 1.65 g, while the biomass was 1.95 g in an agitated liquid conical flask, and in growtek bioreactor the callus attained a mass weight of 2.11 g after 45 days of incubation. Veerabathini et al. (2015) found that MS + 1.0 mg/l BAP + 1.0 mg/l NAA + 1.0 mg/l casein hydrolysate + 0.6% agar was the best media for the growth (85% response) and development of both callus and friable callus from leaf explants of *C. roseus* after 16 days of cultivation. Rashmi and Trivedi (2015) reported that 2,4-D (2.5 mg/l) was the best for callus induction from leaf and stem explants of *C. roseus* (85% in stem and 87% in leaf). Day of callus induction started from 13th to 37th day. This variation is due to the differences in culture conditions and the age of explants. Khashan and Husain (2015) reported that the highest callus fresh weight (3.267 g) was found when leaf explants from *C. roseus* were grown on MS medium containing 1.0 mg/l 2,4-D + 1.0 mg/l BA.

Vermicompost extract along with the coelomic fluid (vermiwash) extracted from the body cavity of earthworm *Eudrilus eugeniae* was tested for in vitro culture and suspension cell culture of *C. roseus* (Kashyab and Kale 2015). In this study callus induction was observed in MS medium after 6 weeks on supplementing the hormones. Vermicompost extract medium alone gave 100% response with respect to callus formation and development of roots and shoots from the callus within 3–4 weeks. Suspension culture was successfully developed using vermicompost extract and coelomic fluid in the 3:1 ratio. Similarly spraying of coelomic fluid (vermiwash) also resulted in 90% callus development, and in both cases the developed callus could be subcultured for four times. Also, Moghe et al. (2016)

found that nodal segment from *C. roseus* mature plants gave the best response when cultured on MS medium supplemented with BAP (1.0 mg/l) + NAA (2.0 mg/l); however, a total of 60 explants were inoculated in the medium out of which 43 explants were responded to callusing. The callus proliferation initiated within 15–20 days. Khashan and Al-Athary (2016) referred that the highest fresh weight (3.267) g was observed when leaves of *C. roseus* were grown on MS medium with 2,4-D (1.0 mg/l) and BA (1.0 mg/l) after 40 days of cultivation. Addition of sucrose or PEG (abiotic factors) to the culture medium causes reduction in callus fresh and dry weight in all treatments except sucrose 40 g/l treatment which gave increase in fresh and dry weight without any difference significant compared with control treatment.

Recently, Alam et al. (2017) observed that MS medium containing 0.5 mg/l NAA and 1.0 mg/l BAP was the best for hypocotyl explant of in vitro grown *C. roseus*; percentage callus formation was 52. Al-Zuhairi and Ghanm (2017) found that the combination of BAP at 2.0 mg/l with NAA at 1.0 mg/l was the best growth regulator for callus initiation from cotyledons excised after 6 weeks from in vitro grown seedlings. Jasmonic acid and glutamine in the culture medium had a positive effect on fresh and dry weights of callus cultures. Mandagi et al. (2017) obtained callus cultures from leaf explant of *C. roseus* shoots grown in MS medium supplemented with 2 mg/l 2,4-D and 0.2 mg/l kinetin. Callus which was 8 weeks old was subcultured for callus propagation using MS medium with addition of 2 mg/l NAA and 0.2 mg/l kinetin. On the fourth day after subculture, callus changes color from yellowish white to brownish yellow. At the age of 12 weeks, callus was produced with a slightly turbid white color.

More recently, Pliankong et al. (2018) established callus cultures from shoot tips of in vivo grown *C. roseus* plants cultured on MS liquid medium supplemented with 1.5 mg/l 2,4-D. For obtaining cell suspension cultures, 12-week-old subculture, the friable callus was transferred to MS liquid medium supplemented with 1.5 mg/l 2,4-D and 0.5 mg/l kinetin. Mekky et al. (2018) obtained callus cultures from leaf explants of *C. roseus* seedlings when grown on MS medium supplemented with 1.5 mg/l BAP and 1.5 mg/l 2,4-D.

3.2 *Plant Regeneration of* C. roseus

It is well known that plant growth regulators are unique component of plant culture media that regulate in vitro morphogenesis processes such as shoot differentiation and affect accumulation of secondary metabolite synthesis. *C. roseus* has been regenerated in healthy plants successfully from in vitro cultures using various techniques, e.g., micropropagation using existing and adventitious meristems, direct and indirect organogenesis using various explants of young seedlings, and somatic embryogenesis using embryogenic callus or other meristemic tissues. In this regard, Pietrosiuk et al. (2007) reported that the biosynthesis of indole alkaloids occurs within the chloroplasts. Accordingly, the formation of dimeric alkaloids may depend on the degree of chloroplast differentiation in the cells cultivated in vitro.

Several studies have investigated the effects of plant growth regulators, explant types, and cultural conditions for establishment of *C. roseus* shoot cultures. In vitro studies of in vitro regeneration from *C. roseus* were shown in Table 1.2.

Hypocotyl explants of *C. roseus* gave rise to adventitious shoots after infection by *A. rhizogenes* at a frequency of up to 80% when cultured on MS medium supplemented with 31.1 μM BA and 5.4 μM NAA. However, a significant difference in the frequency of adventitious shoot formation for each hairy root line is derived from a different cultivar. The produced plants exhibited prolific rooting and had shortened internodes. Half of the plants had wrinkled leaves and an abundant root mass with extensive lateral branching but otherwise appeared morphologically normal (Choi et al. 2004). It was found that friable embryogenic callus was induced from hypocotyl of *C. roseus* grown on MS medium supplemented with 2,4-D (1.0 mg/l) and only NAA (1.0 mg/l) produced somatic embryos in cultures. Callus showing embryogenesis (73.00%) was on MS + NAA (1.0 mg/l) + BAP (1.5 mg/l). MS maturation medium containing 3% maltose gave the best embryo length (11.47 mm) in 7 weeks. Plantlet conversion was better achieved when embryos were matured on MS + (1.0 mg/l) GA_3, 3% fructose or 3–6% maltose. Complete plant conversion was raised in BAP (0.5 mg/l) containing medium (Junaid et al. 2006). Hypocotyl explants of *C. roseus* were placed on MS medium supplemented with 1.0–2.0 mg/l either of 2,4-D, NAA, or chlorophenoxyacetic acid (CPA) for callus initiation (Junaid et al. 2007). Numerous somatic embryos were induced from primary callus on MS medium supplemented with 1.0 mg/l NAA. For embryo maturation, embryonic callus was transferred to medium with GA_3 (1.0 mg/l). The highest length of somatic embryos 10 and 8.85 mm in solid and liquid medium, respectively, was observed in that medium. Embryo proliferation was much faster on medium supplemented with BAP (1.5 mg/l) + NAA (1.0 mg/l). The highest number of somatic embryos (99.25, 2.27 per 40–50 mg embryogenic callus) was obtained on medium supplemented with 1.5 mg/l BAP + 1.0 mg/l NAA. Mature green embryos were developed and germinated well into plantlets with roots on MS liquid medium supplemented with 0.5 mg/l BAP. Best shoot lengths 18.80 and 12.50 mm were observed in liquid medium and in solid medium, respectively, at 0.5 mg/l BAP. Before transfer ex vitro, plantlets were cultivated on half-strength MS medium containing a 0.5 mg/l BAP for additional 2 weeks, and then the plantlets were transferred to outdoor.

Micropropagation could be a good alternative for the mass propagation of *Vinca minor*. Single-node explants of field-grown *Vinca minor* were cultured on a medium consisting of WPM salts, MS vitamins, 3% sucrose, 0.8% agar, and different combinations of BAP and NAA. The results showed that the maximum shoot regeneration (5.6 shoots per explant) was obtained using 7.21 mg/l BAP and 0.0186 mg/l NAA. The average shoot length was 1.36 cm, and some of the explants produced roots. The concentration of 1 mg/l NAA resulted in the highest average root length (3.85 mm). The regenerated shoots were then planted in plastic cups, filled with a mixture of peat moss/perlite/sand (3:1:1) and kept in a growth chamber at 25 ± 2 °C under a 16 h photoperiod (Raouf Fard et al. 2008). However, Swanberg and Dai (2008) reported that cultivar Pacific Coral showed the maximum regeneration rate (73.3%) when internodal explants were placed on WPM containing 5 μM BA and

Table 1.2 Studies on in vitro regeneration of *C. roseus*

Explant type	Culture condition	Establishment stage	Multiplication stage	Rooting stage	Reference
		Response			
Hairy root from transgenic hypocotyl	Cultures were maintained under light 16 h photoperiod at 25 °C	MS; NAA (5.4 μM) + BA (31.1 μM)	Hairy roots are capable of producing adventitious shoots that subsequently developed into mature plants		Choi et al. (2004)
Hypocotyl-derived callus	Cultures were incubated under a 16 h photoperiod at 25 °C light/20 °C dark temperature	MS + NAA (1.0 mg/l) + BAP (1.5 mg/l)	MS + (1.0 mg/l) GA_3, 3% fructose or 3–6% maltose	MS; BAP (0.5 mg/l)	Junaid et al. (2006)
Hypocotyls of 5- to 7-day-old germinated seeds (Primary callus initiation)	Cultures were incubated under a 16 h photoperiod and temperature of 25 ± 1 °C	MS; NAA (1.0 mg/l) (Embryogenic callus initiation) MS; GA_3 (1.0 mg/l) (Maturation of embryos)	MS; BAP (1.5 mg/l) + NAA (1.0 mg/l) either on solid or in liquid proliferation medium	½ MS; BAP (0.5 mg/l)	Junaid et al. (2007)
Single-node explants of field-grown plants (*Vinca minor* cv. Budakalasz)	All cultures were incubated at 25 ± 2 °C under a 16 h photoperiod	WPM salts + MS vitamins; BAP (7.21 mg/l) + NAA (0.0186 mg/l)		MS; NAA (1 mg/l)	Raouf Fard et al. (2008)
Leaf and internodal explants from in vitro plants	Cultures were placed under a 16/8 h photoperiod at 21 °C	WPM; NAA (5.0 μM) + BA (5.0 μM) (Cultivar: Pacific Coral) WPM; NAA (10 μM) + BA (20 μM) (Cultivar: Sunstorm Rose)		MS; NAA (5.0 μM)	Swanberg and Dai (2008)
Mature zygotic embryo (cv. "little bright eye")	For somatic embryogenesis, the cultures were incubated in the dark. For organogenesis the cultures were placed in a culture room with 10 h dark/14 h light photoperiod All of the cultures were incubated at 25 ± 1 °C and subcultured at 21-day interval	MS; TDZ (7.5 μM) (Somatic embryogenesis) MS; either TDZ (2.5 μM) or MS; NAA (5.3 μM) and BA (2.2 μM) (Organogenesis)		½ MS; IBA (2.2 μM) +1–2% sucrose	Dhandapani et al. (2008)

Hypocotyl and cotyledon		MS; NAA (5.3 μM) + BA (2.2 μM)			
Stem nodes and meristem tips		NAA and BA			
Shoot tip explants from in vitro seedlings (*C. roseus*; Egyptian genotype)	Cultures were incubated in a growth chamber at 25 ± 2 °C under a 16/8 h light dark photoperiod	MS; NAA (1.0 mg/l) + BAP (1.0 mg/l)			Taha et al. (2008)
Shoot tip explants from in vivo seedlings (*C. roseus*; three genotypes)	Cultures were maintained under a cool white fluorescent light	MS; 2,4-D (2.0 μM) + TRIA (5 μM) (Initiation of somatic embryos)	MS; sucrose (3.0%) + ABA (5 μM) + agar (0.8%) (Maturation)	Half-strength MS (Plantlet conversion)	Malabadi et al. (2009)
Petiole explants of field-grown plants	Cultures were maintained in a culture room at 18 -h photoperiod. The room temperature was 25 °C	MS; BA (0.2 mg/l) + NAA (0.75 mg/l)		MS; BA (0.3 mg/l) + NAA (1.0 mg/l)	Zulkepli and Samad (2011)
Hypocotyl explants were excised from 4-day-old germinated seedlings (cultivars Pacifica cherry red, Heatwave mix color, and Mediterranean Rose Red)	All of the cultures were incubated at 25 ± 2 °C during a 16 h photoperiod	MS; 2,4-D (1.0 mg/l) + NAA (1.0 mg/l) + ZT (0.1 mg/l) + CH (150 mg/l) + L proline (250 mg/l) + sucrose (30 g/l) + gelrite (3 g/l) (Primary callus induction)	MS; BA (5.0 mg/l) + NAA (0.5 mg/l) (Somatic embryo germination)	MS; BA (1.75 mg/l) + IAA (0.55 mg/l) (Plantlet formation)	Yuan et al. (2011)
Leaf explants were submerged in a 13%, 15%, or 20% (w/v) mannitol containing cell protoplast washing solution for 0, 15, 30, 60, 120, and 180 min. *C. roseus* (cv. Nirmal, Prabal, and Dhawal)	All stock cultures were incubated at 24 ± 2 °C under a 16 h light and 8 h dark photoperiod	½ MS; BA (7.0 mg/l) + NAA (3.0 mg/l) (Shoot bud morphogenesis)	½ MS for 2–3 weeks MS; BA (1.0 mg/l) + NAA (0.1 mg/l) + 0.4 mg/l thiamine HCl, + 4 g/l phytagel (Shoot elongation)	½ MS; IBA (3.0 mg/l)	Verma and Mathur (2011b)
Shoot tip and axillary node explants from in vivo plants	Cultures were incubated at 25 ± 2 °C under 16 h photoperiod	MS; NAA (4.0 mg/l) + BA (4.0 mg/l)	MS; NAA (4.0 mg/l) + BA (4.0 mg/l)	½ MS; IBA (4.0 mg/l)	Bakrudeen et al. (2011)

(continued)

Table 1.2 (continued)

Explant type	Culture condition	Establishment stage	Multiplication stage	Rooting stage	Reference
		Response			
Shoot tip explants from seedlings grown in greenhouse	Cultures were maintained at 26 °C with 16 h light/8 h darkness or completely dark conditions	MS; BAP (1.5 mg/l) + NAA (1.0 mg/l)	MS; BAP (3.0 mg/l) + NAA (4.0 mg/l)	½ MS; IBA (2.5 mg/l) + NAA (0.5 mg/l)	Singh et al. (2011)
Nodal segment explant from in vivo plants	Cultures were incubated at 25 ± 5 °C with 8 h photoperiods	MS; BAP (0.5 mg/l)	MS; BAP (0.5 mg/l) + NAA (1.0 mg/l) + 3% activated charcoal	½ MS; IBA (0.10 mg/l)	Faheem et al. (2011)
Nodal segments from seedling grown in vitro		MS; BAP (0.5 mg/l) + (NAA (1 mg/l)		½ MS; IBA (0.1 mg/l)	Rahmatzadeh et al. (2014)
Hypocotyl-derived callus	Cultures were incubated under a 16 h photoperiod at 25 °C light/20 °C dark temperature	MS; 2,4-D (1.0 mg/l) (Initiation of somatic embryos)	MS; 2,4-D (1.0 mg/l) or MS; NAA (1.0 mg/l) + BAP (1.5 mg/l) (Maturation of somatic embryos)	(Plantlet conversion)	Aslam et al. (2014)
Embryogenic callus	Cultures were incubated under a 16 h photoperiod at 25 °C light/20 °C dark temperature	MS; NAA (5.37 μM) + BAP (6.62 μM) (Initiation of somatic embryos)	MS; GA_3 (2.60 μM) (Maturation of somatic embryos)	MS; BAP (2.22 μM) (Plantlet conversion)	Mujib et al. (2014)
Shoot tip explants from in vivo plants	Cultures were maintained for 16 h a day at a temperature of 25 ± 1 °C	MS; BA (2.0 mg/l)			Al-Oubaidi and Amin (2014)
Nodal explants obtained from 15-day-old seedlings	Cultures were incubated under temperature (25 ± 2 °C), light (2000–2500 lux for 16 h)	MS; BAP (1.0 mg/l)	MS; BAP (1.0 mg/l) or MS; BAP (0.5 mg/l) + NAA (1.0 mg/l)	¼ MS; IBA (5.0 mg/l)	Begum and Mathur (2014)

Explants stem (nodal) and leaf of were collected from the medicinal plant garden	Calli were incubated under a 16/8 h (light/dark) photoperiod	MS; BAP (1.5 mg/l) + NAA (0.5 mg/l) or MS; BA (2.5 mg/l)		MS; NAA (2.0 mg/l) or MS; BAP (1.5 mg/l) + NAA (0.5 mg/l)	Rashmi and Trivedi (2015)
Hypocotyl- and cotyledon-derived callus tissue *C. roseus* cv. Pacifica pink	Cultures were incubated at 24 °C with a 16 h light, 8 h dark, photoperiod	MS; NAA(1.0 mg/l) + BAP (1.0 mg/l)		MS; IAA (10 mg/l) + BAP (0.1 mg/l) or MS; IBA (10 mg/l) + BAP (0.1 mg/l)	Makhzoum et al. (2015)
Shoot tip explant from mature plants	Cultures were incubated at 16:8 h photoperiod and temperature of 27 ± 20 °C	MS; BAP (2.0 mg/l) + Kin (1.0 mg/l)			Moghe et al. (2016)
Hypocotyl explants from in vitro germinated seedlings	Cultures were maintained at white fluorescent light for 16 h with 25 ± 2 °C	MS; BAP (4.0 mg/l) and NAA (0.05 mg/l)	MS basal; growth regulator-free proline (250 mg/l), casein hydrolysate (150 mg/l) and sucrose (3%)	MS; IBA (1.0 mg/l)	Alam et al. (2017)

WPM salts Lloyd and McCown (1980), *TRIA* triacontanol, *ABA* abscisic acid

5 µM NAA, while cultivar Sunstorm Rose showed a regeneration rate of 56.7% with a combination of 20 µM BA and 10 µM NAA. The regeneration rate decreased when the explant size was greater than 7 mm for Pacific Coral cultivar. For rooting of the produced shoots, the highest rooting rate was 100% for the shoots grown in MS medium with 5.0 µM NAA and produced roots that were shorter and thicker than normal roots. Rooted plants were transplanted in potting mix and acclimatized under ambient condition. An average of 60% of the rooted plants survived in the greenhouse and bloomed in 2–3 months.

Dhandapani et al. (2008) showed that the highest regeneration percentage through somatic embryogenesis was obtained from mature zygotic embryo explant of *C. roseus* grown on MS medium supplemented with 7.5 µM of thidiazuron (TDZ). In this medium it was found that a number of somatic embryos formed and mature embryos percentage were 14.7 and 48.7, respectively. The mature embryo also regenerated efficiently via organogenesis in MS medium supplemented with either 2.5 µM TDZ or 5.3 µM NAA and 2.2 µM (BA). Hypocotyl and cotyledon did not induce somatic embryogenesis and organogenesis in TDZ-containing medium but gave a maximum percentage of shoots in MS medium supplemented with 5.3 µM NAA and 2.2 µM BA. Stem nodes and meristem tips showed better regeneration via organogenesis in the medium supplemented with NAA and BA and in lower concentrations of TDZ. Maximum rooting efficiency in regenerated shoots (100%) was observed in half-MS medium containing IBA (2.2 µM) + 2% sucrose. Rooted shoots were transferred to sterile pot mixtures, acclimatized in the culture room, and transferred to a greenhouse (Fig. 1.3). Taha et al. (2008) found that the best medium for direct shootlet regeneration from different explants of Egyptian *C. roseus* was recorded with MS supplemented with 1.0 mg/l each of NAA and BAP. Regarding the type of explant, it was found that the highest number of induced shoot regeneration (189) was recorded with shoot tip explants. In another study, shoot tips from in vivo plants of *C. roseus* cultured on MS medium supplemented with 2.0 µM 2,4-D and 5 µM TRIA were the best for producing the embryogenic tissue. The embryogenic tissue was subcultured on maturation medium (MS medium supplemented with 3.0% sucrose, 5 µM ABA, and 0.8% agar) for further development of somatic embryos. The highest percentage of somatic embryogenesis (85.0) was recorded in genotype III with a total of 14 somatic seedlings recovered per gram fresh weight of embryogenic tissue. Somatic embryos were successfully germinated on half-strength MS medium without growth regulators. Plantlets were placed in a growth room under a 16 h photoperiod for hardening (Malabadi et al. 2009).

Young leaves, stems, and petioles from mature plants of *C. roseus* were cultured on MS medium with different concentrations of NAA and BA for shoot organogenesis. Zulkepli and Samad (2011) found that shoots were regenerated from petiole explant from mature plants of *C. roseus* cultured on MS medium supplemented with 0.2 mg/l BA and 0.75 mg/l NAA. As for leaves and stems, explants did not induce shoot formation in medium containing BA and NAA. However, the highest number of root per explants (0.8) was regenerated from leaf explants cultured on MS medium supplemented with 0.3 mg/l BA and 1.0 mg/l NAA.

Fig. 1.3 Adventitious shoot regeneration from six explants of *C. roseus* cv. "little bright eye" in MS medium supplemented with 2.2 μM BA and 5.3 μM NAA. Shoot regeneration from the callus initiated from the cut end of mature zygotic embryo (**a**). Shoot regeneration from the callus induced from the cut end of etiolated hypocotyl after 40 days of culture (**b**). Shoot regeneration from the callus induced from the cut end of cotyledon after 40 days of culture (**c**). Direct shoot formation from the petiole (**d**). Shoot regeneration from the callus induced from the shoot tip after 40 days of culture (**e**). Shoot regeneration from the callus induced from the stem node after 40 days of culture (**f**). Root formation from the regenerated plant in 1 = 2 MS medium supplemented with 2% sucrose and 2.4 μM IBA (**g**). Fully grown plant showing flowers (**h**). Simultaneous emergence of somatic embryo and adventitious shoot from mature embryo in MS medium supplemented with 2.5 μM TDZ (**i**). *Sh* adventitious shoots, *Se* somatic embryos. (500 × 375). (Adapted from Dhandapani et al. 2008)

Yuan et al. (2011) found that hypocotyl explants of three *C. roseus* cultivars produced the highest percentage of calluses (100) on MS medium supplemented with 1.0 mg/l 2,4-D + 1.0 mg/l NAA and 0.1 mg/l ZT. MS medium with 5.0 mg/l BA and 0.5 mg/l NAA was the most effective medium in inducing and proliferating somatic embryos within 2 weeks of culture. Somatic embryos were germinated well into plantlets on medium supplemented with 1.75 mg/l BA and 0.55 mg/l IAA, within

6 weeks of culture. It could be observed that the genotype Pacifica cherry red showed the best formation of primary, embryogenic callus and regenerated plants than other genotypes. Plantlets were transplanted to potting soil and maintained in a growth chamber. Verma and Mathur (2011b) found that a 60 min pre-plasmolytic treatment of leaf explants from *C. roseus* in a cell protoplast washing (CPW) medium containing 13% (w/v) mannitol followed by their plating on a half-strength (MS) medium supplemented with BA (7.0 mg/l) and NAA (3.0 mg/l) resulted in the de novo induction and development of adventitious shoot buds in more than 75% of explants. The exposure of explants to CPW 13% mannitol stress also helped in reducing the time period for shoot bud emergence to 30–35 days in comparison to 60–70 days in non-treated controls. For shoot elongation and plantlet development, cultures were transferred to a half-strength basal medium for 2–3 weeks and then to full-strength MSM medium containing 1.0 mg/l BA and 0.1 mg/l NAA + 0.4 mg/l thiamine HCl and 4 g/l phytagel. The rooting in the regenerated shoots was obtained in the presence of IBA (3.0 mg/l), and the rooted plants could be established in soil with a 70% rate of success.

Bakrudeen et al. (2011) showed that the highest number of shoots (19.6 shoots/auxiliary node) was observed after 45 days of culture axillary bud and shoot tip explants of *C. roseus* in the MS medium supplemented with NAA (4.0 mg/l) + BA (4.0 mg/l). Shoots were proliferated and elongated in the same medium. High frequency of rooting (82.5%) and high root length (3.65 cm) was obtained in half-strength MS + 4.0 mg/l IBA from axillary bud-derived shoots. The rooted plantlets were successfully established in soil. Also, Singh et al. (2011) found that the highest shoot regeneration in cultures of *C. roseus* was observed as percent response and number of shoots per shoot tip explant (88.9 and 15–17) on media containing BAP (1.5 mg/l) + NAA (1.0 mg/l). Besides maximum shoot regeneration, response (72.4%) was observed from hypocotyl calli both under light and dark conditions on media supplemented with BAP (1.5 mg/l) + NAA (1.0 mg/l). For rooting of the produced shoots, half-strength MS medium supplemented with IBA (2.5 mg/l) + NAA (0.5 mg/l) gave best rooting response (90%) with quality roots. The regenerated plantlets were transferred to pots, and the plantlets grew vigorously in the net house. Faheem et al. (2011) reported that the highest number of shoots per nodal segment explant of *C. roseus* (9.2) was found on MS medium with 0.5 mg/l BAP with shoot length, 3.4 cm, within 7 weeks of inoculation. Maximum multiple shoots (35.10) and shoot length (6.56 cm) were produced on MS medium supplemented with BAP (0.5 mg/l) + NAA (1.0 mg/l). Best rooting response (93.0%) was obtained on half-strength MS containing IBA (0.10 mg/l) where 30.05 healthy roots/shoots were produced with 6.87 cm of root length after 4 weeks of culture. Regenerated plantlets were successfully acclimatized and hardened off inside the culture room and then transferred to greenhouse with 100% survival rate.

Tryptophan has an indirect role on the growth via its influence on auxin synthesis which contributes in promoting plant growth and development (Abou Dahab and Abd El-Aziz 2006). In this respect, Rahmatzadeh et al. (2014) showed that MS medium containing BAP (0.5 mg/l) + NAA (1.0 mg/l) was the best medium for regeneration from growing nodal segment explants of *C. roseus*. It was also recorded that the number of shoot per explant was 5.75. Half-MS medium supplemented with IBA (0.1 mg/l) was the best for rooting of shoots (86%). However, adding tryptophan (250 and 350 mg/l) into this optimal medium resulted in the maximum shooting and rooting percentages, respectively. In another study, hypocotyl explants of *C. roseus* were excised and placed on MS medium with 1.0 mg/l 2,4-D for induction of callus. The callus later transformed into embryogenic callus medium containing with the same concentration of 2,4-D (1.0 mg/l) or in medium amended with NAA (1.0 mg/l) and BAP (1.5 mg/l). The embryogenic callus was white, granular, friable, and fast growing and started to differentiate somatic embryos within 4–6 weeks of culture. The advanced somatic embryos appeared to have well-organized shoot meristem with root primordial ends, which later turned into leaves and roots, respectively (Aslam et al. 2014). Moreover, Mujib et al. (2014) found that the highest number of somatic embryo (99.75) was recorded with agar-solidified MS medium containing NAA (5.37 μM) and BAP (6.62 μM) after seventh week of culture. Embryo length was also increased on MS liquid medium in agitated conical flask containing 2.60 μM GA_3, and after 4 weeks the length reached up to 9.83 mm. The best somatic embryo germination (plantlet conversion) percentage (59.41) was recorded on MS medium with BAP 2.22 μM in growtek bioreactor. Root and shoot length were 7.69 and 11.25 mm, respectively. However, the recovery time for obtaining somatic seedlings from embryogenic callus on solid medium was 135–150 days. The plantlet recovery period was reduced to a minimum of 115 days in liquid medium, and in growtek bioreactor the period further lowered to 112 days. The regenerated plants were healthy and grew normally in outdoor conditions.

Al-Oubaidi and Amin (2014) found that the highest shoot number (4.75), leaves number (9.25), fresh weight (1103.75 mg), and dry weight (112.00 mg) were recorded when shoot tips from *C. roseus* plants grown on MS medium supplemented with 2.0 mg/l BA. Also, Begum and Mathur (2014) obtained shoots from nodal explants of *C. roseus* when grown on MS medium containing 1.0 mg/l BAP and subcultured on the same fresh medium after every 21 days for multiplication. The best response was observed on medium containing 0.5 mg/l BAP + 1.0 mg/l NAA with average number of shoots (7.30) and average shoot length (5.97 cm). However, the best rooting response (90%) was observed on 1/4 MS medium containing 5.0 mg/l IBA, where the number of roots was 3.60 and root length was 1.68 cm. The in vitro rooted plants were hardened first under controlled conditions of culture room and then shifted to mist house where they exhibited, hence, growth and 90% survival. Rashmi and Trivedi (2015) established callus cultures from leaf and stem explants of *C. roseus*. For regeneration of shoots from callus tissue, the MS medium with BAP (2.5 mg/l) was the best for shooting (79%). In double combinations, BAP (1.5 mg/l) and NAA (0.5 mg/l) were the best for shooting (85%). Day of shoot generation started from 19th to 39th day. Higher induction of root (91%) was observed at NAA (2.0 mg/l) after 12 days of cultivation, and in double combinations, BAP (1.5 mg/l) and NAA (0.5 mg/l) were the best for rooting (89%) after 9 days of cultivation. Regenerated plants after hardenings were transferred to soil, and they showed 77% survival. Makhzoum et al. (2015) reported that shoots were obtained from hypocotyl- and cotyledon-derived callus tissue of *C. roseus* after 1–2 months on MS medium supplemented with 1.0 mg/l each of NAA and BAP. Shoot formation was highest (70%) from callus derived from hypocotyl explants after 35 days of culture. Shoots were subsequently rooted on medium supplemented with either IAA (10 mg/l) and BAP (0.1 mg/l) or IBA (10 mg/l) and BAP (0.1 mg/l), acclimated and transferred to the greenhouse, where they developed and produced flowers. In addition, shoots were readily obtained in high yield from apical meristems cultured on MS medium supplemented with NAA and BAP both at 1.0 mg/l. The highest efficiency achieved (97%) was with the cultivar Mediterranean Lilas. Moghe et al. (2016) found that in the medium containing MS + BAP (2.0 mg/l) + Kin (1.0 mg/l), 246 *C. roseus* shoot tip explants from 450 inoculated explants responded to multiple shoot induction with regeneration percentage of 54.57. Recently, Alam et al. (2017) found that MS medium with BAP (4.0 mg/l) and NAA (0.05 mg/l) evoked maximum response of 35.0 (80%) shoots per hypocotyl explant of *C. roseus* after 5 weeks of culture period. For elongation and multiplication, the clusters of developed shoots were excised and cultured on growth regulator-free MS medium (MS basal along with proline (250 mg/l), casein hydrolysate (150 mg/l), and sucrose (3%)). The microshoots were excised from the shoot clusters and cultured on rooting medium. IBA proved to be the best at 1.0 mg/l concentration, and recorded 30.0 roots per shoot were produced in 100% cultures in 5 weeks of culture period.

3.3 *Cryopreservation of* C. roseus

It was reported in many published works that the efficiency of cell lines in culture often deteriorated significantly with time (subculture times). The superior cell lines capable of high production of alkaloids require preserving to maintain their capacity for long time. The critical steps in cryopreserving plant cells cultured in vitro are the time of sampling or period of growth phase, nature and type of cryoprotectants employed, treatment with cryoprotectants, cooling rates, and the terminal freezing temperature before the specimen is stored in liquid nitrogen (Kartha 1981). The work of cryopreservation of in vitro cultures from *C. roseus* by various authors is summarized in Table 1.3.

A procedure for prolonged cryogenic storage of periwinkle cell cultures is described (Kartha et al. 1982). Cells derived from *C. roseus* and subcultured as suspension in B5 medium have been frozen, stored in liquid nitrogen (−196 °C) for 11 weeks, thawed, and recultured. The results showed that maximal survival was achieved when 3–4-day-old cells precultured for 24 h in medium with 5% DMSO were frozen at slow cooling rates of 0.5 or 1 °C/min prior to storage in liquid nitrogen. Chen et al. (1984a) reported that cultured *C. roseus* cells treated with DMSO or sorbitol into the liquid growth medium had a significant effect in the temperature range for initiation to completion of ice crystallization. Compared

Table 1.3 Studies on cryopreservation of *C. roseus* cultures

Culture type/condition	Preculture additives	Cryoprotectants Freezing/method Storage regime	Results; comments	Reference
Cell culture derived from anthers (line 916) All callus and suspension cultures were grown in continuous light at 28 °C	3–4-day-old cell cultures were precultured in 5% (DMSO) Adding 5 ml of culture medium containing 5% DMSO over a period of 15 min and placed on a gyratory shaker at 150 rpm for 24 h	The cell suspensions were centrifuged at 100 × g for 3 min Aliquots were distributed into cryogenic glass ampoules and arranged in the freezing chamber The specimens were cooled at a rate of 0.5–5 °C/min to a terminal temperature of −40 °C followed by immersion in liquid nitrogen (−196 °C) The ampoules stored in liquid nitrogen for 1 h were rapidly thawed in a 37 °C water bath for 90 s and transferred immediately to an ice bath The thawed cell suspensions were centrifuged and diluted with six volumes of chilled culture medium over a 30 min period and washed three times by centrifugation The cells were plated onto solid medium containing 0.8% agar or cultured in liquid medium	3–4-day-old subcultures were most suitable for freezing Cells which were precultured for 1 day in culture medium containing 5% DMS0 showed 66% survival or viability loss of 34% occurred subsequent to preculture in DMS0 Maximal viability (100%) was obtained when the cells were frozen at cooling rates of 0.5 or 1.0 °C/min to −40 °C followed by 1 h storage in liquid nitrogen	Kartha et al. (1982)
Cell lines (PRL no. 200 and a cell line which does not produce detectable levels of alkaloids, PRL no. 916) were initiated from anthers of *C. roseus* Cultures were grown on a gyratory shaker (130 rpm) under a continuous light at 24 ± 1 °C	3–5-day-old cultures of cell line no. 200 were transferred to centrifuge tubes and centrifuged at 100 g for 3 min The liquid was removed, and the packed cells were placed over B5 medium with 1 mg/l 2,4-D and 1 g/l casein hydrolyzate, 1 M sorbitol, 0.5% agar	Aliquots of cell suspension were dispensed into cryogenic glass ampoules flamesealed and arranged in the freezing chamber precooled to 0 °C The specimens were then cooled at 1 °C/min to −40 °C followed by immersion in liquid nitrogen After 1 h storage in liquid nitrogen, the ampoules were rapidly thawed in a 37 °C water bath	There was a close association between the percent unfrozen water at −40 °C and percent cell survival after freezing for 1 h in liquid nitrogen Both DMSO and sorbitol reduced the rate of ice crystallization in solutions Sorbitol is effective in preventing water from freezing at temperatures above −25 °C	Chen et al. (1984a)

(continued)

Table 1.3 (continued)

Culture type/condition	Preculture additives	Cryoprotectants Freezing/method Storage regime	Results; comments	Reference
Alkaloid-producing cell lines (PRL nos. 200, 615, and 91,601) were initiated from callus of anther walls and filaments of *C. roseus*	4-day-old suspension of three cell lines was precultured in media with 1 M sorbitol for 6–20 h The cells were then incubated in media with 1 molar sorbitol +5% DMSO in an ice bath for 1 h	The cells were frozen in the previous solution at a cooling rate of 0.5 °C/min to −40 °C prior to immersion in liquid nitrogen (LN) After rapid thawing in a 40 °C water bath, the regrowth of LN stored cells was achieved by transferring them without washing onto filter paper discs over media solidified with agar for a period of 4–5 h The filter paper discs with the cells were then transferred to fresh media of the same composition for regrowth	The combination of 5% DMSO with 0.5 or 1 M sugars or sugar alcohols improved survival of cells One hour preculture in 1 M sorbitol improved the rate of cell survival from 8% to about 40%, and the rate of survival reached a peak of 60% within 6 h The viability was about 80%, 70%, and 60% for the cells precultured in media enriched with 1.0, 1.25, and 1.5 M sorbitol, respectively	Chen et al. (1984b)
Callus cultures originate from anthers of *C. roseus* (cv. Little Delicata) Suspension cultures initiated from callus on liquid B5 medium with1 mg/l 2,4-D, + 0.1 mg/1 kin and 1 g/1 casein hydrolysate Cultures were kept at 24 °C in a light/dark regime of 16 h/8 h	Preculturing was for 24 h in 1 M sorbitol in normal growth medium 5% DMSO was added as protectant	Freezing was at 0.5 °C/min to −40 °C whereafter the ampoules were immersed in liquid nitrogen Samples were rapidly thawed in a +40 °C water bath. The thawed material was transferred to solidified normal growth medium on a filter paper disc, and recovery growth was measured *Storage under mineral oil:* Callus was transferred to 25 ml of normal solidified culture medium in a wide-mouthed tube and covered with a 2.5 cm overlay of sterile liquid paraffin. The tubes were kept at 10 °C for 6 months, whereafter the calluses were transferred to normal growth medium for recovery growth	The two different methods of preserving cell cultures maintained the growth behavior similarly Preservation under mineral oil does not preserve the productivity (indole alkaloid) of cell cultures, whereas the cryogenic method does	Mannonen et al. (1990)

Embryogenic cell cultures derived from hypocotyl explant Cultures were maintained under a 16 h photoperiod at 25 ± 2 °C on a rotary shaker at 120 rpm	Preculture medium consisting of 5.37 μM NAA + 6.62 μM BA + sucrose (0.09, 0.2, 0.4, and 0.6 M) or sorbitol (0.2, 0.4, and 0.6 M) for 48 h	Vitrification solution of either DMSO (5% or 10%) or glycerol (5% or 10%) or both at six different levels either alone or in combination Cultures of each treatment were first incubated at −20 °C for 2 h and then transferred to liquid nitrogen for 1 h. Then removed from the liquid nitrogen, thawed rapidly in a water bath at 37 °C for 2 min, and then poured onto a fresh liquid medium consisting of 0.09 M sucrose, NAA and BA for detoxification Cultures were placed on a rotary shaker at 120 rpm for 24 h and then transferred to an optimized solidified MS for regrowth	Embryogenic biomass improved in the presence of 0.4 M sorbitol Sucrose at 0.4 M was noted to be efficient treatment for preserving cellular viability The highest survival rate (69.54%) was observed when cultures were subjected to a medium containing 0.4 M sucrose, 5% DMSO, and 5% glycerol Cultures subjected to 0.4 M sucrose and 5% DMSO were also effective in protecting cells (63.59% survival) Highest regrowth (88.9%) was observed when suspension cultures were subjected to 0.4 M sucrose pretreated with 5% DMSO	Fatima et al. (2009)

to the control, less water crystallized at temperatures below −30 °C in DMSO-treated cells. Sorbitol had less effect on the amount of water crystallized at temperatures below −25 °C. The combination of DMSO and sorbitol was the most effective in preventing water from freezing. The results confirm that a minimal amount of liquid water is essential for a cell to survive freezing; it appears that the amount of liquid water at −40 °C is critical for a successful cryopreservation. In this respect, it should be mentioned that DMSO at high concentrations is known to alter membrane permeability and also to interfere with the RNA and protein synthesis (Kartha et al. 1982).

Chen et al. (1984b) showed that when preculture of cell lines from *C. roseus* was followed by freezing in 5% DMSO + 1 M sorbitol, cells did survive freezing and LN storage. The combination of 5% DMSO and 1 M sorbitol enabled 61.6% of cells to survive the freezing and storage in LN. The frozen (LN) and thawed samples of all three cell lines did grow to the extent of control (no. 200) or lesser than control (nos. 615 and 91,601). For the alkaloid-producing cell lines, the maximal cell survival was achieved by slowly freezing at 0.5 °C/min to −40 °C prior to immersion in LN. The viability after thawing as evaluated by the 2,3,5-triphenyl tetrazolium chloride method was about 60% of controls. However, the alkaloid identification showed that the cryopreserved cells retained their alkaloid-producing capability. Selected constituents of the alkaloid spectra of cell lines (200, 615, and 91,601), i.e., corynanthe- and aspidosperma-type alkaloids, were detected prior to and after cryopreservation, and the total alkaloid content of cultures remained basically unchanged. In another work, the growth behavior and productivity of *C. roseus* cell line were studied after 6 months of preservation in liquid nitrogen and under mineral oil (Mannonen et al. 1990). The results showed that neither growth kinetics nor the degree of vacuolization during growth was affected by the long-term preservation. After a recovery period and compared with cultures maintained by frequent subcultivation during the same period, the amount of indole alkaloids in cell culture decreased to 1–2% of the original during 12-month cultivation. On the other hand, preservation with mineral oil did not preserve the productivity of indole alkaloids. In contrast, it was reported that the storage under mineral oil has also been very useful for the pre-cells for in vitro preservation which resumed growth in renewed suspensions on fresh liquid media (Bachiri et al. 1995). Moreover, Fatima et al. (2009) revealed that the cryopreservation treatment combinations of either 0.4 M sucrose, 5% DMSO, and 5% glycerol or 0.4 M sucrose and 5% DMSO resulted in the highest frequency of viability of cell suspension cultures derived from hypocotyl explant of *C. roseus*. However, the treatment (5% DMSO) produced the highest number of cell colonies (10.06) following reculture of cryopreserved cultures. All calluses regrown in an optimized medium, containing 6.62 μM (BA) and 5.37 μM (NAA), resumed normal growth and produced somatic embryos similar to those from nonfrozen embryogenic cultures. These somatic embryos were converted into regenerated plantlets, and all plantlets exhibited normal morphology. Cryopreservation of embryogenic cells and plant regeneration was shown in Fig. 1.4.

Fig. 1.4 Cryopreservation of embryogenic cells and plant regeneration in *Catharanthus roseus*. (**a**) Cryopreserved embryogenic suspended cells produced colonies on plated solid medium. (**b**) Induced callus (arrow, top left) and embryos (arrow, top right and bottom left) from cryopreserved culture. (**c**) Mature somatic embryo with shoot and root axis (arrow). (**d**) Plant regenerated from cryopreserved embryogenic cell. (Adapted from Fatima et al. 2009)

3.4 Genetic Transformation of C. roseus

3.4.1 *Agrobacterium tumefaciens*-Mediated Transformation

Genetic transformation of *C. roseus* represents a real challenge due, in part, to the lack of regeneration capability and to the recalcitrant nature of this species to genetic transformation (Makhzoum et al. 2015). Chosen and recent studies on *A. tumefaciens*-mediated transformation of *C. roseus* are presented in Table 1.4.

Highest transformation efficiency of 1.4 transgenic shoots/responded explant was obtained when pre-plasmolyzed leaves of *C. roseus* were preincubated on shoot bud induction medium for 10 days and subjected to sonication for 30 s prior to transformation. Using a selection medium containing 50 mg/l kanamycin, transformants grew into microshoots and then transferred to half-strength MS basal medium elongated into normal shoots, which formed roots on this hormone-free medium. The transformed plants have broader and more numbers of leaves under in vitro conditions. The transgenic nature of the regenerated plants was confirmed by PCR amplification of uidA gene and GUS histochemical assay (Verma and

Table 1.4 *A. tumefaciens*-mediated transformation studies on *C. roseus*

Infected explant	*Agrobacterium* strain/gene construct	Results/response	Reference
Leaf explants were excised from multiple shoot cultures of Dhawal variety	LBA4404 harboring a binary vector pBI121 with p35SGUS-INT gene construct	Sonication treatments of preincubated explants for 60 s gave highest regeneration and transformation frequency Co-cultivation phase of 120 h resulted in highest transformation frequency in terms of GUS-positive regenerants with minimum callusing response The GUS-positive shoots were also analyzed by PCR amplification of transgene uidA. The 366 bp expected uidA fragment was present in the positive plasmid control as well as in all analyzed shoots	Verma and Mathur (2011a)
Hypocotyls (cultivar Pacifica cherry red)	EHA105 harboring a binary vector pCAMBIA2301 containing a report β-glucuronidase (GUS) gene and a selectable marker neomycin phosphotransferase II gene (NTPII)	The highest frequency of *GUS* transient expression (100%) and the relatively lower death rate (5%) were obtained for the explants with 2-day co-cultivation and 100 μM of acetosyringone in co-culture medium in dark The southern results indicate that the T-DNA was inserted into genome of both transgenic plants The expression level of *DAT* was significantly increased by 7.64-fold in some transgenic plants. The real-time PCR suggests that *DAT* expression is associated with the accumulation of vindoline	Wang et al. (2012)
Meristem tissues (9- to 10-day-old seedlings)	AGL1 harboring the binary vector pCAMBIA 1305.1	Density of bacterial colony had an impact on gene expression, with cultures adjusted to an $OD_{600} = 0.5$ resulting in higher GUS activity than with cultures used at a higher optical density The duration of the co-cultivation period had an effect on gene expression since those co-cultured for 4 days showed a higher level of GUS activity than for those co-cultivated for only 2 days Transformation of meristem tissues with *A. rhizogenes* resulted in a typical hairy root phenotype, in which the adventitious root tissue strongly expressed the p35S *GUSPlus* construct, as revealed by intense GUS staining	Makhzoum et al. (2015)
Cell suspension lines raised from a hairy root clone (PVG) of *V. minor* with integration of all the three genes *rolA*, *rolB*, and *rolC* was used	LBA1119 having TDC + STR (<hpt <Tdc2-<Str-gus>) construct The construct contains a binary vector pMOG22 encoding resistance to the hygromycin and having a version of the GUS reporter gene	Sonication exposure (30 s) leads to successful overexpression of TDC + STR genes The incorporation of construct TDC + STR in the cell suspensions has also been confirmed by PCR-based GUS gene detection. All the three lines were found to be positive for GUS and rol gene-specific PCR products Maximum of twofold increase in the total alkaloids was found in PVG3 (2.1% dry wt) in comparison to control cell suspensions (1.14% dry wt)	Verma et al. (2015)

Hypocotyl obtained from 6-day-old seedlings	LBA4404 harboring GFP gene construct cloned into pGreen0029 (pRepGFP0029) tagged with newly identified strong constitutive promoter	Pre-culturing explants for 6 days on MS basal medium with proline (250 mg/l) and casein hydrolysate (150 mg/l) were prerequisites for the efficient transformation (76%) and help in regeneration of explants infected with the *Agrobacterium* Co-cultivation period was assayed in the 2 days in dark at 27 °C which leads to higher rates of transformation All the GFP-positive transformants showed the presence of 500 bps nptII gene as a selectable marker gene	Alam et al. (2017)
Hypocotyl and nodal explants	Agrobacteria strain GV3101 harboring pBI121::G(G)PPS and pBI121::GES overexpression constructs	The transformation efficiency using nodal explants was 3% similar to transformation using hypocotyl explants The various transgenic lines showed increased expression of G(G)PPS with a 2- to 9.5-fold increase compared to WT control plants Transgenic plants co-expressing G(G)PPSCGES showed increased expression of both genes HPLC analysis revealed that the trend in accumulation of monomers correlated with the level of secologanin in all seven lines and G(G)PPSCGES co-expressing lines with significant increase alkaloids in T1 G(G)PPS and G(G)PPSCGES lines and T1 G(G)PPS lines exhibited improved accumulation of monomeric alkaloids with no significant change in levels of secologanin	Kumar et al. (2018)
Cell suspension cultures (cell-line MP183L)	The pTH2-Δ*EcoRI* and pRT101 plasmids were used Plasmids containing *CrGES* and Δpl*CrGES* were introduced into cells via biolistic transformation	Northern blot analysis and the reverse transcriptase PCR confirmed the overexpression of the *CrGES* (plastid) and Δpl*CrGES* (cytosol) in some of the transformed cell suspension cultures Neither TIA nor iridoids were detected in the transformed or control cell cultures. Only tryptophan and tryptamine were detected in all transformed cell lines, similar to the non-TIA producing cell line Jasmonic acid elicitation stimulated the TIA production in most transformed cell lines, but did not cause clear differences in TIA production between overexpression and control cultures	Saiman et al. (2018)
Leaf-derived callus (cv. Dhawal)	LBA1119 strain having tryptophan decarboxylase and strictosidine synthase gene cassette	In callus and leaf transient transformation, only 36-h-grown bacterial suspension (OD600 = 0.50) with 48 h co-cultivation duration provided the positive GUS staining Transformed callus and leaf tissues with *CrTDC* and *CrSTR* gene overexpression resulted in enhanced relative expression of *CrTDC*, *CrSTR*, *CrSGD*, *CrDAT*, and *CrPRX* transcripts	Sharma et al. (2018)

Mathur 2011a). An efficient *Agrobacterium*-mediated transformation system for *C. roseus* plants was reported (Wang et al. 2012). The results indicated that the transformation frequency reached at 11%. The sonication treatment of 80 W and 10 min can improve the transformation efficiency, keeping relatively high surviving rate (85%). However, the expression of *GUS* gene was confirmed by histochemistry, polymerase chain reaction, and genomic southern blot analysis. The HPLC results reveal that the production of vindoline was enhanced in transgenic plants with *DAT* overexpression. The amount of vindoline was 1.42~2.72 μg/mg (DW) in transgenic plants with overexpressed *DAT* and about 1.15 μg/mg (DW) in both pCAMBIA2301 and wild-type plants. The rooted plants grew normally after transplanted to soil in the greenhouse. It took 3~4 months using hypocotyls as starting materials from inoculation with *A. tumefaciens* to transplant to soil. Makhzoum et al. (2015) reported that although meristem-derived shoots of *C. roseus* transformed with *A. tumefaciens* revealed transient *GUS* expression and protein accumulation, they were not amenable to selection even after 2 months on selection medium. A typical hairy root phenotype was obtained when meristems were treated with *A. rhizogenes* harboring the same p35S::*GUS* construct as with *A. tumefaciens*. Adventitious roots obtained in this manner demonstrated intense in situ GUS activity, especially when these were derived from tissue below the main apical meristem.

Verma et al. (2015) found that 30-s SAAT (sonication-assisted *Agrobacterium*-mediated transformation) treatment of plant cell suspension of *V. minor* with LBA1119 generated three stable TDC + STR overexpressing cell lines, PVG1, PVG2, and PVG3. Real-time (RT) qPCR analysis of PVG3 showed more than seven- to eightfold increase in TDC and STR expression in PVG3 cell line, respectively. Maximum biomass production in cell lines overexpressing TDC + STR was acquired on the 30th day in comparison to the control cell suspensions which produce maximum biomass on the 20th day. The PVG3 cell line was also upscaled in the 5 l stirred-tank bioreactor. Only PVG3 line registered a twofold increase in total alkaloid content (2.1% dry wt) and showed vincamine content (0.003% dry wt) which was further enhanced at the bioreactor level (2.7% and 0.005% dry wt, respectively), while control cell suspensions showed total alkaloid content equivalent to shake flask culture. Alam et al. (2017) reported that hypocotyl explants responded best for maximal production of transformed shoots. It was found that *Agrobacterium* cultured for 30 min in MS liquid medium followed by co-cultivation for 48 h resulted in enhancement of the transformation rate (46%) in SAAT as compared to 23% in conventional method. The transformation efficiency of SAAT method was 6.0% comparable to 3.5% as conventional method. The microscopic studies revealed that the expression of GFP was clearly localized in leaf tissue after transformation of pRepGFP0029 construct. PCR analysis confirmed the integration of the nptII gene in the transformed plantlets.

It has been previously reported that the terpene moiety limits MIA biosynthesis in *C. roseus*. To overcome this limitation and enhance the production of monomeric and dimeric MIA levels, bifunctional geranyl(geranyl) diphosphate synthase [G(G)PPS] and geraniol synthase (GES) that provide precursors for early steps of terpene moiety (secologanin) formation were overexpressed transiently by agroinfiltration and stably by transgenic *C. roseus* plants that were generated through *Agrobacterium*-mediated transformation. Kumar et al. (2018) showed that both transient and stable overexpression of G(G)PPS and co-expression of G(G)PPSCGES significantly enhanced the accumulation of secologanin, which in turn elevated the levels of monomeric MIAs. In addition, transgenic plants exhibited increased levels of root alkaloid ajmalicine. The dimeric alkaloid vinblastine was enhanced only in G(G)PPS but not in G(G)PPSCGES transgenic lines. Moreover, first generation (T1) lines exhibited comparable transcript and metabolite levels to that of T0 lines. Transgenic lines displayed normal growth similar to wild-type plants indicating that the bifunctional G(G)PPS enhanced flux toward both primary and secondary metabolism. Previous studies showed that geraniol could be an upstream limiting factor in the monoterpenoid pathway toward the production of terpenoid indole alkaloid (TIA) in *C. roseus* cells and hairy root cultures. This shortage in precursor availability could be due to (1) limited expression of the plastidial *geraniol synthase* resulted in a low activity of the enzyme to catalyze the conversion of geranyl diphosphate to geraniol or (2) the limitation of geraniol transport from plastids to cytosol (Saiman et al. 2018). Therefore, *CrGES* or Δpl*CrGES* genes were constructed into plasmid vectors and transferred into *C. roseus* cells by particle bombardment. In this study, *C. roseus*' geraniol synthase (*CrGES*) gene was overexpressed in either plastids or cytosol of a non-TIA producing *C. roseus* cell line. The expression of *CrGES* in the plastids or cytosol was confirmed, and the constitutive transformation lines were successfully established. The results showed that overexpression of *CrGES* in the plastids or cytosol caused alteration of primary metabolism that is associated to the plant cell growth and development. HPLC analysis showed that the transformed cell lines did not produce TIA or iridoid precursors unless elicited with jasmonic acid, as their parent cell line. This indicates a requirement for expression of additional inducible pathway genes to reach production of TIA in this cell line. More recently, Sharma et al. (2018) showed that transformed calli and leaf of *C. roseus* using *A. tumefaciens* strain having tryptophan decarboxylase and strictosidine synthase gene cassette gave GUS-positive staining and showed the presence of *uidA* gene band in PCR assay. Total alkaloid content in transformed callus line varied from 0.9 to 1.2 mg when compared to 0.6 mg in the non-transformed control tissue, while alkaloid content in transformed leaves varied from 0.8% to 1.2% on a dry weight basis in comparison to 0.8% dry wt in the non-transformed control. The gene expression analysis of transformed leaves revealed that ATL3 and ATL4 lines recorded a maximum of fourfold and a fivefold increase in *CrSTR* and *CrSGD* transcript levels, respectively. The gene expression analysis of *CrTDC*, *CrSTR*, *CrSGD*, *CrDAT*, and *CrPRX* genes in transformed callus lines recorded the maximum *CrTDC*, CrSTR, and *CrSGD* transcripts levels in ASC2 line followed by ASC4, ASC3, and ASC1 lines. The study also reported the accumulation of strictosidine, ajmalicine, catharanthine, serpentine, and tabersonine in *CrSTR* overexpressing cell lines having tenfold higher *CrSTR* activity in comparison to wild-type tissue.

3.4.2 *Agrobacterium rhizogenesis*-Mediated Transformation

It was reported before by many workers that hairy root cultures established by *A. rhizogenes*-mediated transformation are widely used as sources of useful compounds due to their rapid growth, genetic stability, and relatively high production of secondary metabolites. It has also been reported that transformed root lines can be a promising source for a continuous production of secondary metabolites. This production can be lasting over successive generations without the loss of biosynthetic stability. However, plants can be regenerated from hairy root cultures either spontaneously or by transferring roots to a hormone-containing medium (Giri and Narasu 2000). Recent studies in establishment of hairy root cultures from *C. roseus* have been attempted by few workers as shown in Table 1.5.

Hypocotyl explants of *C. roseus* produced hairy roots when cultured on MS basal medium after infection by *A. rhizogenes* (Choi et al. 2004). Callus formed on hairy root explants gave rise to adventitious shoots at a frequency of up to 80%. Excised adventitious shoots were rooted at a frequency of 100% within 2 weeks. Each hairy root line derived from a different cultivar exhibited a significant difference in the frequency of adventitious shoot formation. Plants derived from hairy roots exhibited prolific rooting and had shortened internodes but otherwise appeared morphologically normal. PCR and southern blot analyses revealed that hairy root lines and plants derived from hairy roots retained the Ri TL-DNA. In another study, hairy root cultures of *C. roseus* were established by infecting the leaves with *A. rhizogenes* agropine-type A4 strain (Batra et al. 2004). The agrobacterium induced roots at the infection site in 52% of explants within 2–3 weeks. There was enormous variability in the morphology and growth patterns among individual root clones. The accumulation of the root-specific alkaloids ajmalicine and serpentine was observed in the presence of both the TL-DNA and TR-DNA genes and the TL-DNA gene alone. Zargar et al. (2010) induced callus cultures from hypocotyl explants of *C. roseus* 15-day-old periwinkle seedlings. The calli (8 weeks old) were placed onto half-strength liquid MS medium that had been polluted by bacterial clones of *A. rhizogenes*. The hairy roots had a suitable growth and 8–9 cm long after 8 weeks. For hairy root production, calli after inoculation were transferred onto solid MS basal medium (free hormone) with 400 mg/l cefotaxime and placed on dark conditions. It was found that the mean number of hairy root regenerated during culture period was the highest (50) after 60 days.

Recently, Hanafy et al. (2016) reported that hairy root was achieved by infecting leaf explants with strain K599 harboring p35SGFPGUS+ plasmid. The root initials on the wounded leaf explants proliferated vigorously and elongated more rapidly when they were subcultured on either half-MS or full-strength MS medium. It could be observed that the transgenic roots increased in fresh mass more than 46-fold and 34-fold in liquid either full- or half-strength MS medium than wild type, respectively. PCR analysis confirmed the presence of HPTII, GUS, and GFP genes. In this study, accumulation of alkaloids (i.e., vinblastine, vincristine, and catharanthine) in the transgenic hairy root was observed in cells and also secreted

Table 1.5 Hairy root culture studies on *C. roseus*

Infected explant	*Agrobacterium* strain	Response	Reference
Hypocotyl (10-day-old seedlings) (cvs. Cascade Appleblossom, Cooler Apricot, Cooler Orchid, Cooler Icy Pink, and Cooler Mix Improved)	*A. rhizogenes* RI000 (This bacterial strain has the chromosomal background of *A. tumefaciens* strain C58 into which the Ri plasmid of pRiA4b has been conjugated)	PCR analysis confirmed the presence of the *rolB* and *rolC* genes in hairy roots and plants derived from hairy Southern blot analysis detected one band (6.5 kb) and two bands (6 and 7 kb) from plants derived from the hairy root lines 2N-1 and 20N-1, respectively	Choi et al. (2004)
Leaves from 1-month-old in vitro grown plants (*C. roseus* var. Prabal)	*A. rhizogenes* agropine-type A4 strain	PCR analysis indicated that the transformed hairy root clones had both complete and partial insertion of Ri T-DNA Southern hybridization analysis revealed both single and multiple copies of T-DNA integration in the root clones	Batra et al. (2004)
The calli (8 weeks old) derived from hypocotyl explants	*A. rhizogenes* RI000	Callus cultures maintained on MS basal medium (free-hormones) formed hairy roots on after 10 days of inoculation	Zargar et al. (2010)
Leaf and stem explants were excised from 3-week-old seedlings (Egyptian *C. roseus* cv. alba)	Cucumopine-type *A. rhizogenes* K599 harboring p35SGFPGUS+ plasmid (plasmid contains GFP-intron (catalase)-uidA gene construct; both genes were under the control of (CaMV) 35S promoter). The hygromycin phosphotransferase (HPTII) gene was also located between T-DNA borders	*A. rhizogenes* cell density was adjusted to give an $OD_{600} = 0.8$ for inoculation Leaf explants induced hairy roots at a higher frequency than stem explants when infected with the *Agrobacterium* strain The best results of hairy root growth were observed when grown in full-strength MS liquid medium, while the growth of wild-type roots in the tested media was very low PCR analysis confirmed the presence of *HPTII*, *GUS*, and *GFP* genes	Hanafy et al. (2016)
Buds of 3-week-old micro-propagated shoots *C. roseus* (cv. Cooler lilac and Cooler blush)	*A. rhizogenes* wild-type strains AR15834 and A4 (with OD_{600} nm 0.6–0.8)	Both the *A. rhizogenes* strains used showed a good transformation frequency In case of 3-day co-culture period, the frequency of root induction reached 90% and 100% for Cooler lilac and Cooler blush, respectively The transgenic nature of hairy roots was confirmed by detecting the presence of *rolB* gene by PCR amplification	Benyammi et al. (2016)

in the liquid medium. Moreover, high antimicrobial activity was demonstrated in the extracts of the transgenic hairy roots. In another study, putative transformants were obtained by co-culturing the explants from two cultivars of *C. roseus* with *A. rhizogenes* for 0–3 days. After 12 days of culture, the hairy root started appearing from wounded sites of explants co-cultured for 3 days with both the bacterial strains. The best results were recorded for Cooler lilac cultivar co-cultured for 3 days with strain A4 (an average of 10.27 hairy roots per explant). The strain AR15834 was most efficient for Cooler blush cultivar showing 91% of transformation frequency. For Cooler lilac, the strain A4 was more active in inducing the hairy root (81% of root induction). In terms of biomass, the established hairy root on liquid medium increased in biomass between 2 and 24 times of the initial inoculums, after 20 days of culture. Among three selected lines, the fastest-growing line was LP10, which showed 24.48 times increase in biomass production as compared with control. PCR analysis confirmed the transgenic nature of the three selected hairy root lines (Benyammi et al. 2016). For more detailed information, an additional table illustrated integrated genes, primer sequences, and PCR conditions used for confirmation of transformation in studies of *C. roseus* hairy roots as shown in Table 1.6.

4 Terpenoid Indole Alkaloid Production from In Vitro Culture of *Catharanthus roseus* Through Biotic, Abiotic Elicitation and Precursor Feeding

Elicitation is one of the most effective techniques currently used for improving the biotechnological production of secondary metabolites. Elicitors are compounds that stimulate any type of plant defense, promoting secondary metabolism to protect the cell and the whole plant (Baenas et al. 2014). According to their origin, elicitors can be divided into different types: (a) biotic and (b) abiotic. Abiotic elicitors can be considered as substances of non-biological origin, being predominantly inorganic compounds such as salts or physical factors (Gorelick and Bernstein 2014). Biotic elicitors have biological origin, often originated as a result of fungi, bacteria, virus, or herbivore infections (exogenous elicitors) and in some cases are released from the attacked plant by the action of enzymes of the pathogen (endogenous elicitors) (Angelova et al. 2006). The majority of biotic elicitors are recognized by specific receptors bound to the cell membrane. This stimulus is then transferred to the cell by a signal transduction system, producing changes that ultimately lead to the formation of phytoalexins (Baenas et al. 2014). Exogenous supply of a biosynthetic precursor to culture medium may also increase the yield of the desired product. The concept is based on the idea that any compound which is an intermediate in a secondary metabolite biosynthetic route can increase the yield of the final product.

Table 1.6 Primer sequences and PCR conditions used for confirmation of transformation for *C. roseus* hairy roots

Gene (MW)	Primer sequences	PCR condition	Reference
rolB gene (860 bp)	F- (5′-CTTATGACAAACTCATAGATAAAGGTT-3′) R- (5′-TCGTAACTATCCAACTCACATCAC-3′)	DNA thermal cycler at 94 °C for 5 min followed by 30 cycles of 94 °C for 45 s, 55 °C for 45 s, and 72 °C for 2 min and a final 7-min extension at 72 °C	Choi et al. (2004)
rolC gene (620 bp)	F- (5′-CAACCTGTTTCCTACTTTGTTAAC-3′) R- (5′-AAACAAGTGACACACTCAGCTTCT-3′)		
virC gene (650 bp)	F- (5′-GGCTTCGCCAACCAATTTGGAGAT-3′) R- (5′-TTTTGCTCCTTCAAGGGAGGTGCC-3′)		
rolA and *rolB* genes of the TL-DNA region	F-(5′-ATGGAATTAGCCGGACTAAACG-3′) R-(5′-ATGGATCCCAAATTGCTATTCC-3′)	The cycling conditions consisted of a 1-min denaturation at 94 °C, a 2-min annealing at 55 °C followed by a 3-min extension at 72 °C, and a final extension at 72 °C for 7 min	Batra et al. (2004)
ags gene of the TR-DNA region	F-(5′-CGGAAATTGTGGCTCGTTGTGGAC-3′) R-(5′-AATCGTTCAGAGAGCGTCCGAAGTT-3′)		
GFP gene (641 bp)	F-(5′-GTAAACGGCCACAAGTTCAGCG-3′) R-(5′-TCGTCCATGCCGAGAGTGATCC-3′)	DNA thermal cycler at 5 min at 94 °C; 35 cycles of 30 s at 94 °C, 40 s at 55 °C, and 1 min at 72 °C and a final incubation step for 5 min at 72 °C	Hanafy et al. (2016)
GUS gene (750 bp)	F-(5′-GATGATAGTTACAGAACCGACG-3′) R-(5′-CATTCGGAATCTCCACGTTAC-3′)		
HPTII gene (812 bp)	F-(5′-CGATTTGTGTACGCCCGACAGTC-3′) R-(5′-CGATGTAGGAGGGCGTGGATATG-3′)		
rolB gene	F-(5′-GCG ACA ACG ATT CAA CCA TAT CG-3′) R-(5′-TTT ACT GCA GCA GGC TTC ATG AC-3′)	DNA thermal cycler 94 °C for 4 min, followed by 30 cycles of denaturing at 94 °C for 1 min, annealing at 57 °C for 1 min, and extension at 72 °C for 1 min and ended with a final extension of 72 °C for 7 min	Benyammi et al. (2016)

4.1 *Abiotic Elicitation and TIA Production from In Vitro Cultures of* C. roseus

Molecular studies suggested that elicitors activate signal transduction pathways which in turn stimulate transcriptional control of several defense genes including genes that encode enzymes participating in alkaloid synthesis pathways (Mujib et al. 2012). For the characterization of growth and alkaloid production of elicited *C. roseus* cell and organ cultures, detailed studies are summarized in Tables 1.7, 1.8, 1.9, 1.10, 1.11, 1.12, 1.13, 1.14, 1.15, 1.16, 1.17, 1.18, 1.19, 1.20, 1.21, and 1.22 and discussed below.

4.1.1 Medium Composition and pH

4.1.1.1 Medium Composition

The effect of type of carbon source, inorganic nutrient phosphate, and nitrogen and macronutrients on terpenoid indole alkaloid (TIA) production from different *C. roseus* cultures (cell, callus, shoot, transgenic, and hairy root) have been extensively investigated. Knobloch and Berlin (1980) stated that medium optimization by altering nutrients (organic and inorganic) and plant growth regulators at different stages of *C. roseus* cultures helps in increasing alkaloid production. Therefore, several efforts were made to investigate the cell and tissue cultures of *C. roseus* to establish as an alternative source of TIA production (Sharma et al. 2019). Response of different in vitro cultures of *C. roseus* to media compositions is presented in Table 1.7.

Increasing phosphate concentrations in the medium of cell suspension cultures of *C. roseus* resulted in decreasing in the intracellular tryptamine very rapidly. Also, the activities of tryptophan decarboxylase (TDC) and of phenylalanine ammonia-lyase (PAL) were influenced differently by phosphate (Knobloch and Berlin 1983). The kinetics of tryptophan decarboxylase activity and the accumulations of tryptophan, tryptamine, ajmalicine, and serpentine were different in three types of media (low sucrose and two high sucrose media each of which differed in their mineral and hormonal contents), but no correlation between maximum enzyme activity and alkaloid contents was observed (Mérillon et al. 1986). In another study, Parr et al. (1988) observed that little alkaloid appeared in the growth medium, but transformed roots of *C. roseus* contained a high level of alkaloids, ajmalicine, serpentine, vindolinine, and catharanthine. However, level of ajmalicine was increased at high sucrose concentration, while level of serpentine was increased at low concentration. Transformed hairy root cultures were found to contain approximately 0.05 μg/g dry weight of vinblastine. Toivonen et al. (1991) reported that the maximal growth of C. *roseus* hairy root cultures was obtained by having 77.8 mg NaH_2PO_4, H_2O/l, and 1.311 g KNO_3/l in the medium, whereas the specific production of alkaloids was highest at the lowest levels of all the nutrients studied.

Table 1.7 Response of different in vitro cultures of *C. roseus* to medium composition

Culture type	Abiotic elicitor	Response	Reference
Cell suspension culture	Media devoid of phosphate and other essential growth factors	Rapid accumulation of tryptamine, indole alkaloids, and phenolics	Knobloch and Berlin (1983)
Cell suspension culture (The cells were agitated on a gyratory shaker, at 24 °C, in the dark)	Three types of media	Growth and alkaloid production differed greatly between the three media used	Mérillon et al. (1986)
Hairy root cultures	Level of sucrose and H_2PO_4 in MS medium	Ajmalicine content was higher in medium with 30 g/l sucrose and 0.1 mM H_2PO_{4-} Serpentine content was higher in medium with 70 g/l sucrose and 0.1 mM H_2PO_{4-}	Parr et al. (1988)
Hairy root cultures (The roots were cultured in the dark on a gyratory shaker (60 rpm) at 24 °C) (cv. Little Delicata)	Sucrose, phosphate, nitrate, and ammonia concentrations	The maximal growth was obtained by having 77.8 mg NaH_2PO_4, H_2O/l, and 1.311 g KNO_3/l in the medium Production of alkaloids was highest at the lowest levels of all the nutrients	Toivonen et al. (1991)
Hairy root cultures (Cultures were maintained in the dark at 25 °C on a gyratory shaker 100 rpm) (cv. "little bright eye")	Sucrose, glucose, and fructose in culture medium	Sucrose, resulting in about 40% (dry wt) higher growth rate Fructose enhanced the catharanthine yield about twofold	Jung et al. (1992)
Cell suspension culture (at 24 °C on an orbital shaker 100 rpm, in the dark)	Effects of diluting mineral salts	Diluting the mineral salts of the culture medium decreased the alkaloid production	Decendit et al. (1992)
	Effects of adding mannitol	Mannitol given alone had no effect on alkaloid production	
	Effects of high levels of $CaCl_2$	$CaCl_2$ (25 mM) gave an ajmalicine yield three times higher than that of control cells	
Compact callus cultures (CCCs) were induced from leaf explants of greenhouse-grown plants	Mannitol (250 mM) and KCl (4 g/l)	Yielded ajmalicine fourfold higher than the control	Zhao et al. (2001a)

(continued)

Table 1.7 (continued)

Culture type	Abiotic elicitor	Response	Reference
Compact callus clusters (CCCs) were induced from leaf explants of greenhouse-grown plants	Sucrose levels varying from 20 to 80 g/l in the maintenance medium	The optimal sucrose concentration for alkaloid and biomass production by the CCC cultures was 50 or 60 g/l	Zhao et al. (2001b)
Suspension culture of the leaves (From naturally grown plants)	Addition of glucose	Adjustment of the osmotic pressure in the medium, light irradiation and addition of glucose (10 g/l) were effective to promote the production of ajmalicine and serpentine	Iwase et al. (2005)
Leaf callus (V. Nirmal)	Strength of the MS media and various levels of carbon sources	Half-strength MS medium supplemented with 0.5 mg/l of 2,4-D and 1.0 mg/l of BA and 6% sucrose was best for biomass production and production of alkaloid	Verma et al. (2012)
Hairy root cultures (*C. roseus*, cv. alba)	Strength of the MS media	The best results of hairy roots growth were observed when grown in full-strength MS liquid medium. However, the growth of wild-type roots in the tested media was very low	Hanafy et al. (2016)
		Amount of vincristine in hairy roots grown in Liquid full MS medium was 1474.3-fold increase than that of control (wild type) Highest accumulation of catharanthine was in the hairy root cultures which were grown on liquid half-MS medium	
Callus cultures	Sucrose (40, 50, and 60 g/l)	Maximum vincristine and vinblastine accumulation were 16.250 and 32.330 µg/g dw, respectively, with 60 g/l sucrose	Khashan and Al-Athary (2016)

Hairy root cultures (Grown on a gyratory shaker rotating at 70 rpm with16/8 h L/D photoperiod at 25 ± 1 °C in shake flask)	Medium composition (the statistically optimized medium recipe was as follows: 2.75 g/l of potassium nitrate,37.5 mg/l of sodium hydrogen monophosphate along with Gamborg B5/2 major salts + B5/2 minor salts + B5 vitamins + 16 g/l sucrose)	Maximum ajmalicine and biomass accumulation were 24.9 mg/l and 5.02 g/l, respectively, on day 30 with practically no unconverted sucrose at the end of the cultivation period Serpentine concentration was (12.3 mg/l)	Thakore et al. (2017)
Callus cultures induced from leaf explants of greenhouse-grown plants	MS and B5 (half strength, full strength, and one and half strength) in agar and in suspension Three different combinations of PGRs Sucrose (1%, 3%, and 5%)	B5 suspension + 0.5 mg/l 2,4-D + 1.0 mg/l kin + 2.0 mg/l NAA + 3% sucrose gave the highest alkaloid content	Mishra et al. (2018a)
Callus cultures induced from leaf explants of greenhouse-grown plants	Glucose, lactose, and maltose were added separately at 1%, 3%, and 5% along with 3% sucrose in both agar and suspension medium	3% glucose added along with 3% sucrose and other standard PGR's combinations stimulated the alkaloid production from the cell culture in suspension medium	Mishra et al. (2018b)
Shoot cultures	KCl (2, 3, 4, and 5 g/l) Mannitol (100, 200, 300, and 400 mg/l)	KCl at 2 g/l levels of supplementation caused more than two- to fourfold increase in growth and biomass accumulation The three elicitation treatments of KCl did not differ much in terms of alkaloid content	Sharma et al. (2019)
Callus cultures		KCl at 3 g/l levels and mannitol at 100 mM supplementation induced 2-fold and 1.7-fold increase in growth and biomass accumulation Elicitation treatments of KCl produced alkaloid content ranging from 1.53% to 2.07% dry wt at 2, 3, 4, and 5 g/l levels A lower dose of mannitol also affects the production of vindoline	

ABT 1-aminobenzotriazole, *CLOT* clotrimazole, *PCA* 2,5-pyridinedicarboxylic acid, *TDC* tryptophan decarboxylase, *PUF* polyurethane foam

The study by Jung et al. (1992) showed that fructose enhanced the catharanthine yield about twofold in hairy root cultures of *C. roseus*. However a two-stage culture using sucrose (1st) and fructose (2nd) markedly increased volumetric yields of catharanthine about twofold (41 mg/l) than that (20 mg/l) of one-stage culture. Diluting the mineral salts of the culture medium decreased the alkaloid production. However, addition of high levels of Ca^{2+}, Mg^{2+}, or Sr^{2+} to B5 media, in which the mineral salts were diluted to 5–40%, increased the alkaloid production (Decendit et al. 1992). Additions of KCl, mannitol, to compact callus cluster of *C. roseus* markedly improved indole alkaloid production and release of these alkaloids into the medium. Treatment with 250 mM mannitol and 4 g/l KCl yielded 42.3 mg/l and 33.6 mg/l of ajmalicine, respectively; these amounts were about fourfold higher than the control (Zhao et al. 2001a). Also, Zhao et al. (2001b) found that sucrose regime affected some properties (the size, degree of compaction, differentiation level) of the compact callus cluster cultures derived from leaf explant of *C. roseus* and therefore influenced alkaloid production. The optimal sucrose concentration for alkaloid and biomass production by the CCC cultures was 50 or 60 g/l.

Adjustment of the osmotic pressure in the medium, light irradiation, and addition of glucose were effective to promote the production of TIAs in suspension culture of the intact plant leaves of *C. roseus*. By feeding glucose (10 g/l) on day 10 of the culture period, ajmalicine was produced at a concentration of about 18 mg/l, and serpentine was produced at a concentration about 11-fold that of the control (Iwase et al. 2005). In contrast with the previous study, Mujib et al. (2012) reported that vitamins and organic components had a low effect on the production of TIAs in most cultures of *C. roseus*.

With regard to effect of medium strength on biomass and alkaloid production of *C. roseus* cultures, Verma et al. (2012) found that half-strength MS basal medium supplemented with 2,4-D and BA (0.5 mg/l and 1.0 mg/l, respectively) and 6% sucrose was best for biomass production of *C. roseus* leaf callus and enhancement of alkaloid accumulation. Hanafy et al. (2016) found that the maximum biomass of *C. roseus* transgenic hairy roots was observed when the hairy roots were cultured in liquid full-strength MS. The transgenic roots increased in fresh mass more than 46-fold and 34-fold in liquid either full- or half-strength MS medium, than wild type, respectively. However, amount of vincristine was 442.3 ng/mg FW in transgenic hairy roots grown in liquid full MS medium and 0.3 ng/mg FW in wild-type roots. Catharanthine in the transgenic hairy roots ranged 0.7 ng/mg FW when grown on liquid half-MS medium. In addition, analysis of the liquid MS medium showed that catharanthine was the highest (17 ng/ml liquid culture) as compared with the other two alkaloids.

Khashan and Al-Athary (2016) showed that the addition of sucrose 60 g/l to the culture medium of *C. roseus* callus tissue resulted in 19.83- and 9.28-fold increase for vinblastine and vincristine, respectively. However, all treatments increase total soluble carbohydrates, proline content, and catalase enzyme activity with significant difference compared with control treatment. Thakore et al. (2017) reported that higher *C. roseus* hairy root biomass was observed on day 30 of cultivation. The ajmalicine production was found to be independent of growth and reduced drastically beyond day 30. No increase was observed in the serpentine concentration

(12.3 mg/l on day 30 to 10.9 mg/l on day 40) which is produced from ajmalicine in the biosynthetic pathway. In another study, various treatments like strength of nutrient salts, sucrose concentrations, and combinations of plant growth regulators were applied to both MS and B5 in agar as well as suspension medium to test their effects on *C. roseus* callus biomass and yield of alkaloid (Mishra et al. 2018a). There was no significant difference in any of the observable parameters when the culture was treated similarly in both types of media formulations (MS or B5). However, the alkaloid content and yield were 2–3 times higher in suspension culture compared to agar medium in similar treatments. The highest alkaloid content observed was 5.67 mg/g dwt in B5 suspension medium containing 3% sucrose and supplemented with 0.5 mg/l 2,4-D + 1.0 mg/l (kin) + 2.0 mg/l NAA.

Alkaloids are being high in carbon containing molecules; it is hypothesized that inclusion of additional carbohydrates along with the preferred sucrose as the carbon sources in the cell culture system of *C. roseus* might enhance its production (Mishra et al. 2018b). It was observed that all three sugars (glucose, lactose, and maltose) tested on biomass development and total alkaloid content of *C. roseus* callus cultures significantly affect various parameters (fresh wt, dry wt, alkaloid content, production, and the yield) in both agar solidified as well as in suspension medium. Although fresh wt and dry wt of cell biomass were 2–3 times less in the suspension medium, the total alkaloid production was 2–3 times higher as compared to agar-solidified medium. However, addition of glucose at 3% concentration along with sucrose at 3% and standard PGR's combinations showed the maximally enhanced total alkaloid yield (0.615% dry wt) as compared to the control (0.605% dry wt). Both lactose and maltose at 3% increased the fresh wt and dry wt, but the total alkaloid production was lesser in comparison to the control cultures. More recently, Sharma et al. (2019) showed that most positive effect on shoot growth was evident in shoot cultures of *C. roseus* treated with low levels of KCl (2 g/l). The highest accumulation (twofold) of catharanthine (0.0729%) was observed in shoots exposed to the highest concentration of KCl (5 g/l). However, gene expression analysis showed that levels of *CrSGD* gene were 1.32-fold increase with KCl (500 mg/l) treatment. With regard to callus cultures, it was found that most positive effect on callus growth was evident in cultures treated with KCl and mannitol. KCl at its lower dose of 3 g/l recorded 12-fold increase (0.0451% dry wt) in vandoline production. However, mannitol at 100 mM dose affects the vindoline production by threefold (0.0115% dry wt) with increased cell growth. Increase in mannitol concentrations reduced the vindoline production. The real-time PCR gene expression analysis of the elicited callus cultures reported that KCl (2.0 g/l) treatment increased the *CrDAT* transcript levels up to 1.34-fold increase. Mannitol (100 mg/l) treatment also recorded 1.20-fold enhanced *CrDAT* transcript levels.

4.1.1.2 Medium pH

pH is an important parameter for controlling growth of plant cell cultures, but growth differs according to differences in plant species, media, and buffer type used. Non-metabolizable buffers can be used to effectively regulate the pH of the

Table 1.8 Response of different in vitro cultures of *C. roseus* to medium pH

Culture type	Abiotic elicitor	Response	Reference
Hairy root clone LBE-6-1 (Cultures were grown in the dark at 26 °C on a rotary shaker 110 rpm)	Different initial pH values of 4.2, 5.7, 6.5, and 7.3	Maximum growth was obtained for cultures with an initial medium pH 6.5 An external pH of 4.2 resulted in abnormal root morphology Sucrose was hydrolyzed completely to glucose and fructose by day 26 except for cultures at initial pH of 7.3	Ho and Shanks (1992)
Hairy root cultures	Influence of buffered media (pH) MES[a] (50 mM) or ACES[b] (20 mM) buffers	Buffers had slightly negative effects upon growth Ajmalicine, serpentine, and horhammericine contents remained unchanged with addition of buffers	Morgan et al. (2000)

[a]MES = 2-(N-morpholino) ethanesulfonic acid – pH range of 6.1 ± 7.5
[b]ACES = 2-[(2-amino-2-oxoethyl) amino]ethanesulfonic acid – pH range 5.5 ± 6.7

media. Furthermore, by controlling pH, the accumulation patterns and amounts of secondary metabolites can be controlled. It seems from the previous published papers that pH value in the culture medium may sometimes fluctuate, affecting the accumulation of alkaloids in *C. roseus* cultures. It may be advantageous to use such a strategy not only to maximize products of interest but also to reduce undesirable products. Few studies have investigated the effects of pH on cell growth and production of TIAs in different *C. roseus* cultures. Response of different in vitro cultures of *C. roseus* to medium pH is presented in Table 1.8.

It was found that maximum growth was obtained from hairy root cultures of *C. roseus* with an initial medium pH 6.5. The lowest growth rate was found in cultures at initial pH values of 4.2 and 7.3 (Ho and Shanks 1992). Negative effect of buffers upon growth of *C. roseus* hairy root culture was observed (Morgan et al. 2000). The yield of lochnericine was significantly lower in response to the addition of buffers, while tabersonine was significantly higher. In contrast, the yields of ajmalicine, serpentine, and horhammericine remained unchanged.

4.1.2 Explant Type, Subculture Cycles, and Characteristics of Cultures

Response of different in vitro cultures of *C. roseus* to explant type, subculture cycles, and characteristics of cultures is shown in Table 1.9. Taha et al. (2008) established callus and shoot cultures from different explants of *C. roseus*. Quantitative comparison of the total alkaloids between the different cultures referred that the highest percentages of total alkaloid (1.116) were recorded from root-derived callus grown on MS + 1.0 mg/l 2,4-D + 1.0 mg/l Kin. However,

Table 1.9 Response of different in vitro cultures of *C. roseus* to explant type, subculture cycles, and characteristics of cultures

Culture type	Abiotic elicitor	Response	Reference
Callus and shoot cultures	Different explant types	The highest alkaloid was observed with root-derived callus and shootlets derived from shoot tip explants	Taha et al. (2008)
Hairy root cultures (LBE 6–1) (The cultures were grown on an orbital shaker at 100 rpm at 26 °C in the dark)	Effect of subculture cycle (Cultures were harvested at 21, 28, and 35 days)	The 2-week subculture cycle yielded the fastest, while the 4-week cycle yielded the slowest growth rates Tabersonine decreased from day 21 to 35, and lochnericine were highest in the 2-week cycle Serpentine yields were lowest in the 2-week cycle and highest in the 4-week cycle	Rijhwani and Shanks (1998)
Compact callus clusters (CCCs) were induced from leaf explants of greenhouse-grown plants	CCCs, friable half-closed hollow callus clusters and dispersed cell suspensions	The CCC cultures produced more indole alkaloids than dispersed cell cultures	Zhao et al. (2001b)

regenerated shoolets derived from shoot tip explants and grown on MS + 1.0 mg/l NAA + 1.0 mg/l BA gave the highest percentage of total alkaloid (1.977). Concerning subculture cycles, it was shown that the subculture cycle affected both growth and alkaloid production. The 2-week subculture cycle *of C. roseus* hairy root culture yielded the fastest growth rates, while the 4-week cycle yielded the lowest specific growth rates. Yields of tabersonine decreased from day 21 to 35, while lochnericine content was highest in the 2-week cycle, and serpentine content was lowest in the 2-week cycle and highest in the 4-week cycle (Rijhwani and Shanks 1998). However, it was reported that, despite maintaining the *C. roseus* cell lines of the transgenic character, the cell lines gradually lost the ability to accumulate terpenoid indole alkaloids (TIAs). The diversity of alkaloids produced was also negatively affected by long-term subculture (Whitmer et al. 2003). Moreover, different types of suspension cultures of *C. roseus* were compared for alkaloid accumulation (Zhao et al. 2001b). The results showed that the compact callus clusters (CCC) could synthesize indole alkaloids 1.9- and 2.4-fold higher than the half-closed hollow callus clusters and dispersed cell cultures, respectively. The total alkaloid production (sum of ajmalicine, serpentine, and catharanthine), or each indole alkaloid content, in the CCC cultures was much higher than that in dispersed cell cultures. Large compact solid callus aggregates gave the highest total alkaloid yield (39 mg/l) than other types of cultures.

4.1.3 Biomass Density and Growth

For the production of secondary metabolites by plant cell cultures of *C. roseus*, a two-stage batch culture has been widely used. In the first stage, optimal conditions for fast growth of biomass are created in growth medium; in the second stage, the secondary metabolite production is induced in induction medium. This volumetric productivity is determined by two factors: the specific productivity and the biomass density.

Response of different in vitro cultures of *C. roseus* to biomass density and growth is shown in Table 1.10. Alkaloid production from *C. roseus* by high cell density culture feeding has been attempted. For biomass density the work by Moreno et al. (1993a) clarified the effect of inoculum density on ajmalicine formation and related enzyme activities in *C. roseus* cells. It was found that in growth medium a transient increase in tryptophan decarboxylase (TDC) activity was first observed after which geraniol-10-hydroxylase (G10H) reached its maximum activity; only tryptamine accumulated, and no ajmalicine could be detected. Low-density culture had accumulated six times more ajmalicine than the high-density culture. Contin et al. (1998a) observed that cultures of higher densities seem to present a limited growth in comparison with those of lower densities. The higher inoculum densities favored terpenoid accumulation, while the lower inoculum densities favored indole accumulation. However, the activities of tryptophan decarboxylase (TDC) were decreased in higher inoculum densities, while higher strictosidine synthase (STR) activity was found for the lowest inoculum densities, which presented higher TDC activities. Recently, Thakore et al. (2017) found that an increase in the inoculum size of *C. roseus* hairy root from 0.5 to 3 g/l (dry weight basis) led to a maximum biomass density at the lowest value. This is a clear indication that with higher initial inoculums, the roots approach senescence faster resulting in reduced biosynthetic capacity as well.

Comparative analysis has been carried out on three selected lines of *C. roseus* hairy roots for their kinetics of growth and the accumulation of ajmalicine and catharanthine, throughout a 35-day culture cycle (Benyammi et al. 2016). Maximum accumulation of the alkaloids is recorded for LP10 line in which the peak of ajmalicine and catharanthine accumulation reached to 3.8 and 4.3 mg/g dry weight (DW), respectively. This increase coincides with an exponential growth phase. The results suggest that the evolution of accumulation of ajmalicine and catharanthine is positively correlated with the development of the biomass growth.

4.1.4 Growth Regulators

The role of plant growth regulators in alkaloid production of *C. roseus* cultures has been extensively studied by different workers. The quantity and quality of growth regulators greatly affected the accumulation of alkaloid in different cultures. In this respect, Moreno et al. (1995) reported that auxin and cytokinin added to the culture

Table 1.10 Response of different in vitro cultures of *C. roseus* to biomass density and growth

Culture type	Abiotic elicitor	Response	Reference
Cell suspension culture (The culture was grown in 1000 ml flasks on a gyratory shaker (100 rpm), at 25 °C in the dark)	Low-inoculum density (5 g FW in 50 ml medium) and high-inoculum density (20 g FW in 50 ml medium)	Increase in biomass concentration (high-density cultures) did not enhance the total alkaloid production	Moreno et al. (1993a)
		Increase in TDC activity was first observed after which G10H reached its maximum activity; only tryptamine accumulated	
		After 30 days the low-density culture had accumulated 6 times more ajmalicine than the high-density culture	
Cell suspension culture (Cultures were placed on a gyratory shaker (80 rpm) at 25 ± 1 °C in continuous light)	Changing the inoculum size (40, 100, 160, 220, 280, 340, and 400 g/l) Cells were harvested at 5, 10, 15, and 20 days after inoculation	The increase of inoculum densities from 40 to 160 g/l caused a significant increase in secologanin content between days 10 and 20 Tryptamine and tryptophan accumulated only in cells inoculated at the lowest density (40 g/l) Ajmalicine and strictosidine accumulation between day 5 and 10 was increased by the increase of inoculum density	Contin et al. (1998a)
Hairy root cultures (Grown on a gyratory shaker rotating at 70 rpm with16/8 h L/D photoperiod at 25 ± 1 °C in shake flask and different bioreactors) (cv. Pra-bal, clone CP 32)	Inoculum density (0.5, 1.25, 2.5, 3 g/l)	Ajmalicine accumulation was maximum (0.44 mg/g) for initial inoculums of 0.5 g/l	Thakore et al. (2017)
Hairy root cultures (Three lines, LP10, LP21, and L54) *C. roseus* (cv. Cooler lilac and Cooler blush)	Biomass growth	There is a linear correlation between the kinetics of biomass growth, and accumulation of ajmalicine and catharanthine LP10 line was the most promising line to continue optimizing the production of TIAs	Benyammi et al. (2016)

media either alone or in combinations routinely regulate alkaloid biosynthesis pathway. Response of different in vitro *cultures of C. roseus* to growth regulators is presented in Table 1.11.

It was found that peroxidases play an important role in the biosynthesis of alkaloids. Addition of 2,4-D to the culture medium reduced the peroxidase activity in the medium and cellular compartments. In particular, the synthesis of a basic isoform was reduced. Cytokinins seemed to affect the acidic isoforms (Limam et al. 1998). Low-benzyladenine (0.1 mg/l) or a benzyladenine-free medium was found to stimulate the morphological development of the multiple shoot cultures of *C. roseus* and increased the production of vindoline and catharanthine; however, growth and ajmalicine production were suppressed. Vindoline and catharanthine were mostly found in leaf tissue, and ajmalicine was localized in the unorganized tissue (Hirata et al. 1987). Cytokinins were found to stimulate the alkaloid synthesis of *C. roseus* cell cultures induced by removing auxin from the medium (Decendit et al. 1992). Garnier et al. (1996) found that treating *C. roseus* callus cultures (generated from untransformed cotyledons) with cytokinins enhanced their alkaloid production while transgenic tissues overproducing cytokinins do not increase levels of indole alkaloids. However, the non-transgenic callus grown without 2,4-D gave the highest content of ajmalicine and serpentine 39.2 and 6824.3 μg/g DW, respectively. Canel et al. (1998) reported that alkaloid accumulation by highly productive transgenic *C. roseus* overexpressing the enzymes strictosidine synthase (STR) and tryptophan decarboxylase (TDC) was reduced in the presence of NAA while the presence of 2,4-D completely eliminated alkaloid accumulation.

Cytokinin and/or ethylene has an effective role in regulation of several genes of indole alkaloid biosynthesis in *C. roseus* suspension cultures (Papon et al. 2005). Addition of cytokinin and ethylene in the suspension cultures of *C. roseus* increased the expression of three genes belonging to methylerythritol pathway. In addition cells grown in medium with auxin did not accumulate any TIAs. Pandiangan and Nainggolan (2006) examined the enhancement of catharanthine content in *C. roseus* callus cultures added with different concentrations of naphthaleneacetic acid (NAA). The results showed that fresh and dry weight of callus increased as the NAA concentration increases. The catharanthine content of callus cultures was increased by adding NAA as well. The highest catharanthine content was found with 2.5 mg/l NAA. Also, Senoussi et al. (2007) found that BAP in medium of *C. roseus* suspension culture increased the ajmalicine production and the effect of hypoxia was completely absent in the presence of cytokinin. However, it was found that the presence of 0.1 mg/l NAA + 0.1 mg/l Kin in the culture medium of *C. roseus* resulted in the highest vindoline, catharanthine, and vincristine contents (Ataei-Azimi et al. 2008). In this study, 10-fold catharanthine, 125-fold serpentine, 0.5-fold vindoline, and 0.34-fold ajmalicine were produced by new roots. It was found that vinblastine and vincristine with amounts of 20-fold and 6-fold, respectively, were increased than that in the petioles of intact plants. Also, NAA (1.0 μM) and BA (0.5 μM) in the culture medium of *C. roseus* callus cultures gave the highest content of vincristine, 20.38 mg/g DW. The maximum amount of dry biomass (598.04 mg) was produced

Table 1.11 Response of different in vitro cultures of *C. roseus* to growth regulators

Culture type	Abiotic elicitor	Response	Reference
Multiple shoot cultures	Different levels of growth regulator	Low or free BA in the medium increased the production of vindoline and catharanthine	Hirata et al. (1987)
Cell suspension culture (at 24 °C on an orbital shaker 100 rpm, in the dark)	Effects of cytokinins	Cytokinins were found to stimulate the alkaloid synthesis	Decendit et al. (1992)
Transgenic callus cultures (cv. Little Pinkie)	Endogenous cytokinins	Alkaloid contents were not increased in the *ipt* transgenic tissues in spite of the high level of endogenous cytokinins	Garnier et al. (1996)
Callus cultures initiated from cotyledons	BA (1 and 5 μM) and zeatin (5 μM)	Exogenously applied cytokinins increased the alkaloid amounts in the callus cultures obtained from cotyledons	
Transgenic cell suspension cultures (overexpress the enzymes strictosidine synthase (STR) and tryptophan decarboxylase (TDC))	Hormonal composition of the culture medium	Alkaloid accumulation by highly productive transgenic lines showed considerable instability and was strongly influenced by culture conditions, such as the hormonal composition of the medium	Canel et al. (1998)
Cell suspensions (C20D) (Incubated at 25 °C on a gyratory shaker at 100 rpm)	Cytokinin and/or ethylene	Enhanced TIAs in cells grown on medium without 2,4-D	Papon et al. (2005)
Callus cultures	Different levels of NAA	The highest catharanthine content was found in callus grown on medium with 2.5 mg/l NAA	Pandiangan and Nainggolan (2006)
Cell suspension cultures (line C20D) on a rotary shaker (100 rpm) at 24 °C in the dark	BAP	BAP can decrease the effects of the hypoxia and increase the ajmalicine production	Senoussi et al. (2007)
Callus and roots were obtained from petiole explant	Different growth regulator combinations	MS with 0.1 mg/l NAA + 0.1 mg/l Kin had the highest vindoline, catharanthine, vincristine	Ataei-Azimi et al. (2008)
		The indole alkaloid levels of new roots in medium containing 0.1 mg/l NAA + 0.1 mg/l Kin were higher than in petioles of intact plants	
Callus derived from stem explants of mature plants	Different growth regulator combinations	MS medium with NAA (1.0 μM) and BA (0.5 μM) gave the highest content of vincristine	Kalidass et al. (2010)

(continued)

Table 1.11 (continued)

Culture type	Abiotic elicitor	Response	Reference
		MS medium supplemented with 2,4-D (1.0 μM) and kinetin (1.0 μM) was the best for callus growth and gave the maximum amount of dry biomass	
Leaf callus *C. roseus* (cv. Nirmal)	Combinations of auxins with cytokinins	The highest production of total alkaloid content was found in 0.50 mg/l of 2,4-D and 1.0 mg/l of BA	Verma et al. (2012)

after 7 weeks of culture (Kalidass et al. 2010). In another study, combinations of auxins with cytokinins were found to be better for *C. roseus* leaf callus growth and enhancement of alkaloid content. The highest content of total alkaloid (96.32 mg/g DW) was found with half-strength MS medium contained 0.5 mg/l of 2,4-D + 1.0 mg/l of BA and 6% sucrose (Verma et al. 2012).

4.1.5 Methyl Jasmonate

Jasmonates (JAs) are plant-specific signaling molecules that activate several important physiological and developmental processes. The biosynthesis of these hormones, which is induced by pathogen attack and wounding, triggers defense responses, both locally and systemically. JAs, particularly methyl jasmonate (MeJa), have been reported to play an important role in signal transduction processes that regulate defense genes in plants (Farmer and Ryan 1990). Response of different in vitro cultures of *C. roseus* to methyl jasmonate is shown in Table 1.12.

Tikhomiroff et al. (2002) found that yields of tabersonine and lochnericine from *C. roseus* hairy roots elicited with 25 mg/l of jasmonic acid were increased by 100–400% and 14–200%, respectively, with the use of silicon oil in non-elicited control cultures. In elicited cultures, these values were 10–55% for tabersonine and 20–65% for lochnericine. In another study, different phytohormones were tested on growth and accumulation of terpenoid indole alkaloids in *C. roseus* cell suspension culture upon feeding with the precursor's loganin and tryptamine (El-Sayed and Verpoorte 2002). Among these only methyl jasmonate induced strictosidine and ajmalicine accumulation. Abscisic acid did not enhance the accumulation of alkaloids but delayed the catabolism of strictosidine. Rodrigueza et al. (2003) reported that tabersonine 6,7-epoxidase activity was detected in total protein extract of jasmonate-induced *C. roseus* hairy root cultures using labeled ^{14}C-tabersonine 2. Shoot clusters of *C. roseus* were grown on liquid medium and accumulated vindoline (Hernandez-Dominguez et al. 2004). Maximal accumulation values (2.0 mg/g dry wt) were reached around day 27 of the culture cycle. Vindoline concentration (2.0 mg/g dry wt) was comparable to that in leaves of intact plants (4.6 mg/g dw). These results suggest that root participation is not required for vindoline synthesis in aerial tissues. However, to shorten the period of vindoline accumulation (around 27 days),

Table 1.12 Response of different in vitro cultures of *C. roseus* to methyl jasmonate

Culture type	Abiotic elicitor	Response	Reference
Hairy root cultures – line LAO (Cultured in a two liquid-phase bioreactor)	Jasmonic acid (25 mg/l) and silicon oil	In elicited cultures, the yields of tabersonine and lochnericine were 10–55% for tabersonine and 20–65% for lochnericine	Tikhomiroff et al. (2002)
		With the use of silicon oil in non-elicited control cultures, the yields of tabersonine and lochnericine were increased by 100–400% and 14–200%, respectively	
Cell suspension culture	Methyl jasmonate	Methyl jasmonate enhanced the accumulation of alkaloids	El-Sayed and Verpoorte (2002)
Hairy root cultures (Line J-1)	Methyl jasmonate	Methyl jasmonate treatment of hairy root cultures significantly activated tabersonine 2 epoxidase activity and increased detectable tabersonine 2 epoxidase activity sixfold within after 48 h of treatment compared to untreated control cultures	Rodrigueza et al. (2003)
Shoots were transferred to MS liquid media with 1.0 mg/l BA and cultured under continuous light on a rotary shaker at 50 rpm	Methyl jasmonate (0.1, 1, or 10 μM)	The highest vindoline content (4.4 mg/g dw) was observed in shoot cultures exposed to 0.1 μM MejA, while control treatment gave 0.4 mg/g dw after 24 h	Hernandez-Dominguez et al. (2004)
Cell suspension cultures (cell line A11) One resin bag (1 g resin) was added to suspensions 3 days after MJ addition	Methyl jasmonate (0, 10, 100, or 1000 μM on either day 0, 3, 6, 9, 12, or 15)	Maximum ajmalicine accumulation (8.1 mg/l) was observed with 100 μM MJ The maximum ajmalicine production (10.2 mg/l) was obtained by day 18	Lee-Parsons et al. (2004)
Inducible expression ORCA3 hairy root line	Elicitation with jasmonic acid	Caused the greatest increase in tabersonine, lochnericine, horhammericine, catharanthine, and serpentine concentrations compared to the uninduced levels	Peebles et al. (2009)
		Induction of ORCA3 overexpression or induction plus elicitation caused a decrease in tabersonine, lochnericine, and horhammericine	

(continued)

Table 1.12 (continued)

Culture type	Abiotic elicitor	Response	Reference
Hairy root cultures (J1 line) (incubated in the dark on a rotating shaker at 100 rpm at 25 °C)	Methyl jasmonate (MeJA) (10, 100, and 250 μM) Root tissue and culture media were collected at 12, 24, 48, and 72 h after elicitation	The highest accumulation (in the roots) of alkaloids was observed after 48 h, when 100 and 250 μM of MeJA were added The highest level of total alkaloid exudation (in the culture medium) was observed after 12 h of treatment	Ruiz-May et al. (2009)
Hairy root cultures	Methyl jasmonate (250 μM MeJA)	The transcript levels of key TIA pathway genes (G10h, Tdc, Str, and Sgd) increased by 9.1-, 3.1-, 6.7-, and 8.3-fold, respectively, after 24 h	Goklany et al. (2009)
		Strictosidine, ajmalicine, and tabersonine levels in MeJA-elicited cultures were 160%, 440%, and 420% increase when compared with controls	
Cell suspension culture derived from calli of leaf explants (cultures were incubated in the dark at 25 °C and shaken at 110 cv. First Kiss Apricot	Methyl jasmonate (MJ) (100 μM)	Ajmalicine production increased when high cell densities (200 g fresh wt/l) were used, in all conditions tested The enhancement of extracellular ajmalicine production is not only dependent on the MJ dose but also on the cell density used and time of elicitation	Almagro et al. (2010)
	MeJA, β-cyclodextrins (β-CDs)[a] (50 mM), and UV light at short and long exposure times (254 nm, 10 μW/cm^2) for 15 or 120 min	A combination of β-CDs and MJ with short exposure to UV light led to a significant increase in ajmalicine production	
Cell suspensions (C20hi)	Methyl jasmonate (MJ) + light	Combining MJ with light enhanced vandoline biosynthesis	He et al. (2011)
Cell suspension cultures were shaken at 80 rpm and maintained in light period of 16 h/day at 23 ± 2 °C	Methyl jasmonate (100 μM) added to T1 cell suspension cultures	The level of ajmalicine in the cells was raised to two- to threefold by MeJA treatment Catharanthine also showed a three- to fourfold higher level than the control	Xiang et al. (2011)
		The ajmalicine content in the medium was 2 mg/l, which was tenfold higher than the control	
Cell suspension culture (Propagated on a rotary shaker (110 rpm) at 25 °C under continuous light)	Jasmonic acid (JA) (101.9 μM/70 ml) Cells harvested at 0, 5, 30, 90, 360, and 1440 min after elicitation	The most significant effects were observed on the accumulation of α-TIA-1 and tabersonine after 360 min of induction, whereas catharanthine and serpentine were induced after 1440 min of elicitation	Goldhaber-Pasillas et al. (2014)

Hairy root cultures (Cultures were incubated at 25 °C with 16/8 light/dark photoperiod on a gyratory shaker at 70 rpm) (cv. Prabal clone CP32)	Methyl jasmonate, jasmonic acid, and KCl (Flasks were harvested after 24 h, 48 h, 72 h, and 96 h)	Increasing ajmalicine production (a fourfold increase compared to batch cultures)	Thakore et al. (2015)
Cambial meristematic cells (CMCs) were cultured in flasks (100, 250, and 500 ml) and in a 5 l stirred hybrid airlift bioreactor	Methyl jasmonate (MeJA) and β-cyclodextrin (β-CD)[a]	Yields of vindoline, catharanthine, and ajmalicine in the bioreactor were higher than those cultured in flasks In the bioreactor and treated with 10 mM β-CD and 150 μM MeJA gave the highest yields of vindoline catharanthine and ajmalicine than yields of CMCs cultured in 100 ml flasks without elicitors	Zhou et al. (2015)
Hairy root cultures (expressed an estrogen-inducible Zct1 hairpin)	Methyl jasmonate (MJ)	With 250 μM MJ, levels of several metabolites increased, including strictosidine, tabersonine, and representative tabersonine-like compounds	Rizvi et al. (2016)
		TIA[b] metabolite levels did not increase further upon Zct1 silencing combined with 250 μM MJ Upon induction with 1000 μM MJ, levels of strictosidine, serpentine, and tabersonine-like compounds were all significantly lower than at 250 μM MJ induction	
Shoot cultures	Methyl jasmonate (MJ) (10, 100, 250 mg/l)	Methyl jasmonate elicitation was found inhibitory for shoot biomass production, 1.5- to 2.0-fold depression over the control was evident in shoots treated with 10 mM and 100 mM methyl jasmonate Methyl jasmonate at higher doses (250 mM) supports minimum alkaloid production in comparison to control	Sharma et al. (2019)
Callus cultures		Methyl jasmonate elicitation was found inhibitory for callus biomass production and showed 1.5- to 2.0-fold depressions over control Higher doses of methyl jasmonate (100–250 mM) support the minimum alkaloid production in comparison to control	

[a]Cyclodextrins (CD) are cyclic oligosaccharides derived from starch with multiple applications in cosmetics and textile, pharmaceutical and food industry due to its interesting properties: increased solubility of nonpolar compounds, increased stability, protection against degradation caused by heat, oxidation, and UV radiation (Singh et al. 2002)

[b]*TIA* terpenoid indole alkaloid

12-day-old shoots were treated with methyl jasmonate. The results showed that exposure to 0.1 μM methyl jasmonate resulted in a tenfold increase within the first 12 h. Despite this increase in vindoline accumulation, tryptophan decarboxylase (TDC) remained at the same activity as in unexposed shoots.

C. roseus cell cultures elicited with methyl jasmonate (MJ) resulted in production of ajmalicine (5.4 mg/l) and serpentine (3.7 mg/l); this represents a 165% and 78% increase over the maximum ajmalicine and serpentine content of controls. However, maximum ajmalicine accumulation of 10.2 mg/l was observed with 100 μM MJ, a 300% increase over the controls (Lee-Parsons et al. 2004). Also, jasmonic acid (JA) treatment caused the largest increase in TIA metabolites and transcripts of pathway genes in *C. roseus* hairy roots. It was found that upon induction of ORCA3 expression and elicitation with JA, ORCA3 transcripts increased 170-fold, whereas ORCA3 expression caused an 89-fold increase and JA elicitation caused a 5-fold increase in ORCA3 transcripts (Peebles et al. 2009). The increase in the concentration of methyl jasmonate (MeJA) *in C. roseus* hairy roots caused an increment in the levels of total alkaloids throughout the temporal course of 12, 24, 48, and 72 h. Moreover, MeJA induced differential secretion of several hairy root metabolites. The secretion of ajmalicine increased significantly in the first 12 h of treatment (Ruiz-May et al. 2009). However, it was shown that gene expression profiling for G10h, Tdc, Str, and Sgd in hairy root cultures elicited with methyl jasmonate (250 μM) increased by 9.1-, 3.1-, 6.7-, and 8.3-fold, respectively, after 24 h. Strictosidine, ajmalicine, and tabersonine levels in MeJA-elicited cultures were 16.2 mg/l, 3.4 mg/l, and 29.3 mg/l, respectively (Goklany et al. 2009).

In recent years, cyclodextrins (CDs) have also attracted considerable attention as agents capable of inducing defense responses in plant cell cultures and therefore acting as true elicitors (Marsh et al. 2014; Perassolo et al. 2016). In this respect, Almagro et al. (2010) described that the joint use of cyclodextrins with methyl jasmonate, when accompanied by a short exposure to UV, enhanced extracellular ajmalicine accumulation to 1040 mg/l in suspension cultured cells of *C. roseus*. However, ajmalicine production slightly decreased as the flask size increased. Similarly, He et al. (2011) reported that the biosynthesis of vindoline was enhanced by combining methyl jasmonate with light in *C. roseus* C20hi cell suspensions, while catharanthine accumulation was not improved by either light or MJ elicitation. In another study, callus and cell cultures from diploid and tetraploid *C. roseus* plant leaves were established (Xiang et al. 2011). It was found that content of ajmalicine reached 2.578 and 0.895 mg/g DW in callus and in cell suspension culture, respectively. However, the cell suspension culture treated with methyl jasmonate increased ajmalicine content to 2.2 mg/g DW in cell and 2 mg/l in culture medium.

Goldhaber-Pasillas et al. (2014) observed late stress response events to cell suspension cultures of *C. roseus* treated with jasmonic acid. They reported that the most significant effects were observed on the accumulation of α-TIA-1 and tabersonine after 360 min of induction, whereas catharanthine and serpentine were induced after 1440 min of elicitation. Increasing feed rate strategy was reported by Thakore et al. (2015) who found that a 60% increase in ajmalicine overall productivity and a 2.5-fold increase in volumetric yield were obtained as opposed to batch cultivation of hairy roots of *C. roseus*. Mixture of jasmonic acid, methyl jasmonate, and KCl was added after the termination of model based increasing fed-batch cultivation which resulted in a significantly high ajmalicine concentration of 123.2 mg/l (a fourfold increase compared to batch).

Elicitation of *C. roseus* cambial meristematic cell cultures in a 5 l stirred hybrid airlift bioreactor with 10 mM β-cyclodextrin and 150 μM methyl jasmonate gave the highest yields of vindoline (7.45 mg/l), catharanthine (1.76 mg/l), and ajmalicine (58.98 mg/l), which were 799%, 654%, and 426% higher, respectively, than yields of cultures grown in 100 ml flasks without elicitors (Zhou et al. 2015). Recently, Rizvi et al. (2016) successfully developed transgenic *C. roseus* hairy root cultures that expressed an estrogen-inducible Zct1 hairpin. Upon induction with 17β-estradiol, Zct1 induction by methyl jasmonate at both low (250 μM) and high (1000 μM) concentrations was considerably reduced. Despite these low levels of Zct1, MJ-elicited TIA production and TIA gene expression (G10h, Tdc, and Str) did not further increase in Zct1-silenced cultures.

Sharma et al. (2019) reported that multiple shoot cultures of *C. roseus* elicited with methyl jasmonate were not found to affect production of catharanthine content. Increase in concentration of methyl jasmonate resulted in a marginal increase in vinblastine production but with a concurrent decrease in the cell growth. Methyl jasmonate (250 mM) treatment showed the presence of trace amount of vincristine. The real-time PCR gene expression analysis of the elicited multiple shoot cultures reported a very slight increase in the levels of *CrSGD* and *CrPRX I* genes. In terms of callus cultures, it was found that the treatments of methyl jasmonate showed a variable decrease in vindoline production compared to non-treated control. Gene expression analysis of the elicited callus cultures failed to enhance the expression of studied genes.

4.1.6 Light and Ultraviolet (UV)-B Light

4.1.6.1 Photoperiod Regime

Light is an important factor for both ex vitro and in vitro morphogenetic study. Its presence, absence, time, and intensity directly influence anabolic and catabolic processes, particularly secondary metabolism. Importance of light to *C. roseus* in vitro cultures was reported. Light can influence enzyme induction and activation. However, light also stimulates plastid development and peroxidase activity and promotes the concentration of vindoline, ajmalicine, and serpentine production in cultured cells, leaves, seedlings, and plants (Zhao et al. 2001a). Autotrophic shoot cultures of *C. roseus* which have the potential of a cost-effective system produce 10% more total alkaloid as compared to mixotrophic cultures (Mitra et al. 1997).

Response of different in vitro cultures of *C. roseus* to photoperiod regime is presented in Table 1.13. Under continuous light conditions, the total amount of TIAs in suspension culture of *C. roseus* was increased about 2.8-fold compared to that in the control (Iwase et al. 2005). Vindoline-producing in vitro shoot cultures were

Table 1.13 Response of different in vitro cultures of *C. roseus* to light, ultraviolet (UV)-B light and gamma radiation

Culture type	Abiotic elicitor	Response	Reference
Cell suspension culture	Continuous light conditions	Total amount of TIAs was increased about 2.8-fold compared to that in the control	Iwase et al. (2005)
In vitro shoot cultures	Continuous light or a 16 h photoperiod regime	Light regime affected biomass gain rather than their morphogenetic potential The highest content of vindoline was with photoperiod at 28 days	Campos-Tamayo et al. (2008)
Cell suspensions C20hi	Light	Light significantly increased vindoline production in C20hi cells by about 0.49- to 5.51-fold more than that in controls	He et al. (2011)
Callus and cell suspension culture derived from leaf explants of both the tetraploid plant (T1) and the diploid control (D1)	Darkness (for callus of T1)	Ajmalicine level reach 2.578 mg/g DW T1 callus synthesized 0.79 mg/g DW ajmalicine, which was about 6.5-fold higher than D1 callus and over 26-fold higher than tetraploid leaves T1 cell suspension culture may be superior to the compact callus cluster suspension culture	Xiang et al. (2011)
Hairy root cultures (Grown on a gyratory shaker rotating at 70 rpm with 16/8 h L/D photoperiod at 25 ± 1 °C in shake flask and different bioreactors) (cv. Pra-bal, clone CP 32)	Illumination period (16/8 h light/dark (L/D) and complete (24 h) dark (D) conditions	Ajmalicine content of 0.39 mg/g was obtained for a photoperiod of 16 h light and 8 h dark	Thakore et al. (2017)
Cell suspensions	Ultraviolet B radiation	UV-B signaling leading to enhanced production of catharanthine	Ramani and Chelliah (2007)
Hairy root cultures (cultures were grown at 26 °C and 100 rpm in the dark)	UV-B light (0, 5, 10, and 20 min of UV-B light) The 24-day-old cultures were exposed to UV-B light for 20 min and harvested 6, 12, 24, 36, 48, 72, and 168 h after exposure	168 h after UV-B exposure shows significant increases in the levels of lochnericine and significant decreases in the levels of horhammericine Increasing UV-B exposure time up to 20 min caused significant increases in lochnericine, serpentine, and ajmalicine and a decrease in horhammericine	Binder et al. (2009)

(continued)

Table 1.13 (continued)

Culture type	Abiotic elicitor	Response	Reference
Regenerated plantlets from seeds grown on MS medium	Gamma radiation and putrescine (grown on control, 60 Gy irradiation + 50 mg/l putrescine and 60 Gy irradiation + 50 mg/l putrescine treatments)	Higher amount of vinblastine was in control treatment and 60 Gy + 50 mg/l putrescine-treated plantlets	Noor-mohammadi et al. (2018)

exposed either to continuous light or a 16 h photoperiod regime (Campos-Tamayo et al. 2008). The accumulation of vindoline in cultures under the photoperiod occurred in coordination with plantlet formation. In treatment with phytoregulators, vindoline synthesis was also reduced via an effect on deacetylvindoline acetyl CoA acetyltransferase (DAT) activity. In another study, it was also noticed that light significantly increased vindoline production in *C. roseus* C20hi cells by about 0.49- to 5.51-fold more than that in controls, with the highest yield being 75.3 ng/g of dry weight (He et al. 2011). It was found that tetraploid callus of *C. roseus* gave the highest content of ajmalicine (2.58 mg/g DW) in the dark condition, while gave the highest content of catharanthine and vindoline, 0.05 and 0.02 mg/g DW, respectively, in the light condition (Xiang et al. 2011). Recently, Thakore et al. (2017) found that *C. roseus* hairy root biomass was higher for incubation in 16/8 h light (L)/dark (D) illumination regime (1.9 g/l) compared to total dark conditions (1.4 g/l). However, ajmalicine content of 0.39 mg/g from hairy root cultures was obtained for a photoperiod of 16 h light and 8 h dark in contrast to 0.16 mg/g for hairy root incubated in complete dark.

4.1.6.2 UV-B Light and Gamma Radiation

In nature, plants generate protective secondary metabolites in response to environmental stresses. Such metabolites include terpenoid indole alkaloids (TIAs), which absorb UV-B light and serve putatively to protect the plant from harmful radiation. UV-B induced alkalinization of the culture medium, generation of hydrogen peroxide, and activation of CDPK (calcium-dependent protein kinases) and MBPK (myelin basic protein kinase). Medium alkalinization, an early event occurring in elicitor-treated plant cell cultures, has been used as a marker of elicitor responses in studying elicitor-binding sites in plant cells (Boller 1995). Response of different in vitro cultures of *C. roseus* to UV-B light and gamma radiation is presented in Table 1.13.

Exposure of *C. roseus* cell suspension culture to low dose of UV-B irradiation was found to increase production of catharanthine and transcription of genes encoding tryptophan decarboxylase (*Tdc*) and strictosidine synthase (*Str*) (Ramani and Chelliah 2007). Analysis of alkaloid concentrations up to 168 h after

UV-B exposure of *C. roseus* hairy roots showed significant increases in the concentrations of lochnericine and significant decreases in the concentration of horhammericine over time. Results also indicate that increasing UV-B exposure time up to 20 min caused significant increases in lochnericine, serpentine, and ajmalicine and a decrease in horhammericine (Binder et al. 2009). Vinblastine production was evaluated in *C. roseus* plantlets grown on MS medium after irradiation by different doses of gamma radiation (Gy) and 50 mg/l putrescine (Noormohammadi et al. 2018). The highest mean value of vinblastine was detected in control leaf samples (233.86 μg/g plant) and 60 Gy + putrescine-treated plantlets (155.88 μg/g plant) than other treatments. However, increase of mean values of morphological traits, *Tdc* and *Str* transcripts, in treated plantlets in comparison to control plantlets was observed.

4.1.7 Temperature and Deprivation of Oxygen

4.1.7.1 Temperature

The temperature is an important element that influences in vitro cultural growth and alkaloid metabolism in *C. roseus*. Various temperatures were tested for their ability to grow and synthesize alkaloids (Bailey and Nicholson 1990). In this respect, Pasquali et al. (1992) reported that the alkaloid metabolism is not only restricted to the specific tissues or cells but is also modulated by different cellular and environmental factors. Response of different in vitro cultures of *C. roseus* to temperature and deprivation of oxygen is presented in Table 1.14.

The temperature is an important element that affects growth and alkaloid production in different cultures of *C. roseus*. Growth rates of cell line of *C. roseus* were maximal at 35 °C but declined rapidly above 35 °C and below 25 °C. The maximum serpentine production was at between 20 and 25 °C. However, ajmalicine showed a sharp peak of accumulation at 20 °C (Morris 1986). In hairy root cultures of *C. roseus*, lowering temperature caused an increase in alkaloid accumulation, but no effect on the distribution of indole alkaloids between the roots and the medium could be detected. The highest content of total alkaloid was observed at 19.5 °C (after 27 days of cultivation) and ranged at 2.56 mg/g DW (Toivonen et al. 1992). The freezing (−196, 4, 15 °C) and nonfreezing (25 °C) temperature was imposed on *C. roseus* embryogenic cultures, and vinblastine production was studied (Junaid et al. 2011). The results showed that somatic embryo production was maximum at 15 °C, but it was high at 4 °C in maturation stage. The highest germination and plantlet conversion were resulted with somatic embryos grown at 25 °C. Maximum vinblastine yield (9.0 μg/g) was noted in germinated embryos in LN-treated cultures. The vinblastine content increased with decreasing incubation temperature. The increased level of vinblastine with decreased temperature might be due to rapid alkaloid metabolism. However, maximum vinblastine contents were extracted from leaf tissues; it was absent in in vitro raised roots.

Table 1.14 Response of different in vitro cultures of *C. roseus* to temperature and deprivation of oxygen

Culture type	Abiotic elicitor	Response	Reference
Cell suspension culture (on an orbital shaker 150 rpm, in the dark)	Various temperatures from 10 to 45 °C	Maximum biomass yields (both fresh wt and dry wt) occurred at 25 °C Maximal rate of serpentine accumulation also occurred at 25 °C Optimal temperature for accumulation of ajmalicine was 20 °C	Morris (1986)
Hairy root cultures (Cultures were incubated on a gyratory shaker (60 rpm) in the dark) (cv. Little Delicata)	Effect of growth temperature (19.5, 24, or 32 °C)	No effect on the distribution of indole alkaloids between the roots and the medium The level of alkaloid accumulation showed a clear increase with lowering temperature	Toivonen et al. (1992)
Embryogenic cultures were incubated at 25, 15, and 4 °C under 16 h photoperiod	Effect of freezing (4, 15, −196 °C) and nonfreezing (25 °C) temperature	Maximum vinblastine was quantified in germinating embryos. The highest vinblastine yield was observed in LN-treated cultures Leaves of in vitro raised plant showed more vinblastine compared with other plant parts	Junaid et al. (2011)
Cell suspension cultures (line C20D) on a rotary shaker (100 rpm) at 24 °C in the dark	Hypoxia (deprivation of oxygen)	Resulted in very strong inhibition in accumulation of the alkaloids with slight reduction of the growth	Senoussi et al. (2007)

4.1.7.2 Deprivation of Oxygen

Senoussi et al. (2007) found that lack of oxygenation in cell cultures of *C. roseus* C20D provokes a very strong inhibition in accumulation of the alkaloids and of other possible substances. Reoxygenation did not restore hypoxia effect on the alkaloid accumulation.

4.1.8 Bioregulators

The close relationship between primary and secondary metabolism, as well as the existence of cross-links between two or more secondary metabolite biosynthesis pathways, led scientists to test several bioregulators with the goal of improving alkaloid production (Godoy-Hernandez and Loyola-Vargas 1991). Bioregulators like phenobarbital, trans-cinnamic acid, succinic acid, and malic acid have been found to enhance the flux into alkaloid pathway, thereby improving the yields (Zhao et al. 2001a). Cytochrome P450 enzymes are widely involved in the biosynthesis of indole alkaloids in *C. roseus* (Meijer et al. 1993), and trans-cinnamic acid blocks the shikimic acid pathway flowing to phenylpropanoid and therefore has been proposed to enhance tryptamine pools for indole alkaloid biosynthesis. Tetramethyl ammonium bromide (TAB) can act as an inhibitor of peroxidase, which widely participates in indole alkaloid metabolism, especially by catalyzing the conversion of ajmalicine to serpentine (Blom et al. 1991). Response of different in vitro cultures of *C. roseus* to bioregulators is presented in Table 1.15.

Morgan and Shanks (1999) treated *C. roseus* hairy root cultures with oxygenase inhibitors, 1-aminobenzotriazole (ABT), clotrimazole (CLOT), and 2,5-pyridinedicarboxylic acid (PCA). They found that ABT specially inhibited the formation of horhammericine, while CLOT inhibited the accumulation of lochnericine. Using jasmonic acid in combination with the inhibitors suggests an inducible P-450 enzyme responsible for the formation of horhammericine. The inhibitor study also revealed that both lochnericine and horhammericine are turned over in hairy root cultures. Zhao et al. (2000) showed that ajmalicine production of cell cultures of *C. roseus* was increased by betaine (to 55 mg/l), *n*-propyl gallate (to 27 mg/l), succinic acid (to 31 mg/l), malic acid (to 60 mg/l), and tetramethyl ammonium bromide (to 64 mg/l). Ajmalicine and catharanthine yields were about five- to sixfold higher than the control. A large portion (up to 50–85%) of total indole alkaloids was released into the medium. Moreover, for maximal catharanthine production, the optimal doses of malic acid and tetramethyl ammonium bromide were 50 mg/l and 120 mg/l, respectively. Godoy-Hernandez et al. (2000) found that application of an osmotic stress and 1 mM trans-cinnamic acid with *C. roseus* cell line cultured in a 14 l bioreactor inhibits the synthesis of phenolic compounds and restored the original alkaloid amount. Also, Zhao et al. (2001a) found that catharanthine production in *trans*-cinnamic acid-treated compact callus cluster cultures of *C. roseus* was 1.5-fold higher than in the control, but phenobarbital had only a limited stimulating effect on catharanthine production. The results also showed that tetramethyl ammonium bromide could significantly improve ajmalicine production (49.3 mg/l) at 60 mg/l TAB-treated cultures (about sixfold higher than that of the control) and catharanthine production (18.3 mg/l) at 30 mg/l TAB-treated cultures.

Gaviraj and Veeresham (2006) showed that treated transformed root cultures of *C. roseus* with phenobarbital increased the accumulation of ajmalicine and serpentine, but catharanthine levels were not affected much when compared with control cultures. As shown in Table 1.15, ajmalicine production (1.98 mg/l) was improved by 1.4-fold over the control cultures (1.44 mg/l) when low concentration of phenobarbital (50 μM) was used. There was 1.5-fold increase in the accumulation of serpentine (2.21 mg/l) when compared with control cultures (1.47 mg/l). Addition of succinic acid resulted in production of ajmalicine (1.86 mg/l) and serpentine (1.97 mg/l) that improved by 1.4-fold each when compared with control cultures (1.31 and 1.45 mg/l, respectively). Catharanthine contents (1.63 mg/l) were improved by 1.2-fold over the control cultures (1.36 mg/l). However, addition of trans-cinnamic acid did not produce significant improvements on alkaloid accumulation. Ramani and Chelliah (2007) found that catharanthine accumulation and transcription of *Tdc* and *Str* in *C. roseus* suspension cultures were inhibited by three- to fourfold upon pretreatment of various inhibitors like suramin, N-acetyl cysteine,

Table 1.15 Response of different in vitro cultures of *C. roseus* to bioregulators

Culture type	Abiotic elicitor	Response	Reference
The hairy root line LBE 6-1 (Cultures were grown in the dark at 26 °C on a rotary shaker 110 rpm)	Oxygenase inhibitors ABT, CLOT, and PCA Jasmonic acid	Addition of inhibitor ABT (1.25 μmol) gave the highest content of tabersonine and lochnericine, while control treatment gave the highest content of horhammericine	Morgan and Shanks (1999)
		The highest content of tabersonine, lochnericine was observed with the addition of JA (0.5 mg) and ABT (0.50 μmol), while the highest content of horhammericine was observed with the addition of JA (0.5 mg)	
Cell suspension culture	Betaine hydrochloride, *n*-propyl gallate, succinic acid, malic acid, and tetramethyl ammonium bromide	Improved ajmalicine, catharanthine, or serpentine biosynthesis	Zhao et al. (2000)
Cell suspension culture	Osmotic stress and trans-cinnamic acid (1 mM)	Inhibits the synthesis of phenolic compounds and restored the original alkaloid amounts	Godoy-Hernandez et al. (2000)
Compact callus cultures	Phenobarbital	Phenobarbital had only a limited stimulating effect on catharanthine production	Zhao et al. (2001a)
	Trans-cinnamic acid	Catharanthine production in *trans*-cinnamic acid-treated CCC cultures was 1.5-fold higher than in the control	
	Tetramethyl ammonium bromide	Improve ajmalicine production (about sixfold higher than that of the control)	
Hairy root cultures	Phenobarbital (50 μM)	Ajmalicine production was improved by 1.4-fold over the control There was 1.5-fold increase in the accumulation of serpentine when compared with control cultures	Gaviraj and Veeresham (2006)
	Trans-cinnamic acid	Did not produce significant improvements on alkaloid accumulation	
	Succinic acid (10 mM)	Ajmalicine and serpentine was improved by 1.4-fold each when compared with control cultures Catharanthine contents were improved by 1.2-fold over the control cultures	

(continued)

Table 1.15 (continued)

Culture type	Abiotic elicitor	Response	Reference
Shoot cultures	Salicylic acid and succinic acid (1, 5, and 10 mM)	Salicylic acid at its lower concentration (1 mM) also increased the production of vinblastine by fivefold with high cell growth increase Most positive effect on shoot growth was evident in cultures treated with higher dose (5–10 mM) of succinic acid	Sharma et al. (2019)
Callus cultures from leaf explants of shoot cultures		The treatments of salicylic acid showed a variable decrease in vindoline production compared to non-treated control Succinic acid with its higher dose of 10 mM resulted in twofold (0.0065%) increase in vindoline content with similar growth compared to non-treated control	

ABT 1-aminobenzotriazole, *CLOT* clotrimazole, *PCA* 2,5-pyridinedicarboxylic acid

inhibitors of calcium fluxes, staurosporine, etc. Recently, Sharma et al. (2019) revealed that salicylic acid supplementation to multiple shoot cultures of *C. roseus* was found dose dependent with a treatment comprising of 10 mM being best with TIA content of 3.72% dry wt. Salicylic acid at its lower concentration of 1 mM also increased the production of catharanthine 1.25-fold, when compared to control. SA (1.0 mM) treatment resulted in the 1.46-fold increased *CrPRX11*. Treatment of succinic acid at 10 mM was best as in addition to 3.15% dry wt alkaloid content; it also supported a fourfold increase in biomass. Also, salicylic acid (5 mM) and succinic acid (1 mM) treatments showed the presence of trace amount of vincristine. With regard to callus cultures, it was found that salicylic acid elicitation was found inhibitory to callus biomass production and showed 2.0- to 3.0-fold and 1.5- to 2.0-fold depressions over control. Alkaloid production in elicitation treatments with salicylic acid ranges from 1.04% to 1.59% dry wt. Highest alkaloid levels of 2.38% and 2.28% dry weight were found in callus fed with 5 mM and 1 mM succinic acid dose and also recorded a 1.6- to 3-fold increase in biomass.

4.1.9 Permeabilizing Agent and Chemicals

Permeabilizing agents are those chemical compounds (solvents) which, under non-inhibitory concentration levels, have the ability to reversibly increase the pore size of the cell wall as well as the vacuoles in order to leach out the metabolites so that the cell grows normally again for better mass transfer with respect to substrate and the product, leading to increased biomass and secondary product formation by plant cells (Thakore et al. 2013). Different permeabilizing agents have been used to facilitate the release of the product stored, including the organic solvents

Table 1.16 Response of different in vitro cultures of *C. roseus* to permeabilizing agent

Culture type	Abiotic elicitor	Response	Reference
Cell suspension culture	In situ adsorption with XAD-7	Enhanced the release of ajmalicine (less than 10% extracellular to 40%) with a 40% increase in total productivity	Asada and Shuler (1989)
Hairy root clones (Cultures were maintained in 250 ml flasks at 25 °C in the dark on a rotary shaker 100 rpm) (cvs. Little Linda- LL and Little Delicata-LD)	Amberlite XAD-2, XAD4, and XAD-7	When XAD-7 was used alkaloid was not detected in the medium A slight increase in growth was observed when XAD-7 was present	Sim et al. (1994)
		The maximum production of catharanthine and ajmalicine was obtained when XAD-7 was added to half-strength SH liquid medium	
	Combined effects of DMSO treatment and in situ adsorption with XAD-7 (0.3 g dry resin/30 ml culture medium)	High DMSO concentrations were not required to release intracellularly stored catharanthine from the hairy roots	
Hairy root cultures (Cultures were incubated at 25 °C with 16/8 h light/dark illumination conditions)	Permeabilizing agents (DMSO, Triton X-100, n-hexadecane, and Tween 80)	Triton X-100 (0.1% v/v) and n-hexadecane (2% v/v) led to 98% increase in the yield of ajmalicine in the roots	Thakore et al. (2013)
		Both DMSO and Triton X-100 were found to be useful in effluxing ajmalicine into the medium resulting in a 12- and 16-fold increase, respectively	

DMSO dimethyl sulfoxide

isopropanol and dimethyl sulfoxide (DMSO) and also chitosan (Brodelius and Nilsson 1983). Dimethyl sulfoxide has been used in many cases, because it is known to extract sterols from the membranes of the eukaryotic cells. The concentration and time of addition of the permeabilizing agent to the medium are key parameters to avoid cell growth inhibition. The presence of a solid adsorbent can be expected to prevent the products from being degraded and to stimulate de novo synthesis by removal of the products. The use of an adsorbent reduces the feedback inhibition by preventing the inhibitory products from being accumulated in the hairy roots. Response of different in vitro cultures of *C. roseus* to permeabilizing agent and chemicals is presented in Table 1.16.

Asada and Shuler (1989) reported that adsorption of ajmalicine from the extracellular medium of *C. roseus* cell culture with Amberlite XAD-7 (a neutral polycarboxylic ester resin) greatly enhanced the release of ajmalicine (less than 10% extracellular to 40%) with a 40% increase in total productivity. Among three Amberlite resins used with *C. roseus* hairy root cultures, XAD-7 greatly enhanced the release of catharanthine and ajmalicine from hairy root cultures, with an increase in total production. Moreover, to enhance alkaloid production and secretion, a permeabilizing agent; dimethyl sulfoxide (DMSO) and a fungal elicitor together with in situ adsorption. Dimethyl sulfoxide (0.5% v/v) treatment with in situ adsorption using XAD-7 was found appropriate for releasing indole alkaloids from hairy roots without affecting cell viability (Sim et al. 1994). Recently, Thakore et al. (2013) showed that both DMSO and Triton X-100 were found to be useful in effluxing ajmalicine from *C. roseus* hairy root cultures into the medium resulting in a 12- and 16-fold increase, respectively. The highest contents of ajmalicine (9.86 mg/g DW) were found in roots when treated with Triton X-100 (0.5% v/v), while the highest content of ajmalicine in the medium was 22.54 mg/l when treated with 0.01%, Triton X-100.

4.1.10 Chemicals

The most actively pursued strategies to increase the production of target natural products in plant cell cultures are the application of chemical elicitors (Zhong 2001). Different workers have investigated the effects of some chemicals as stress factor on stimulating the accumulation of alkaloids in cultures of *C. roseus*. Response of different in vitro cultures of *C. roseus* to some chemicals is presented in Table 1.17.

Nitric oxide (NO) is known as a signaling molecule involved in elicitor-induced defense responses of plants. Sodium nitroprusside (SNP), a donor of NO, stimulates catharanthine formation in *C. roseus* cells. Addition of 0.1 mM SNP did not affect the growth of *C. roseus* hairy roots, whereas 1 mM SNP suppressed the growth significantly, and 10 mM SNP almost completely inhibited the growth of the hairy roots. It was also found that SNP treatment stimulated the formation of serpentine, catharanthine, ajmalicine, horhammericine, lochnericine, and tabersonine by day 21 (Li et al. 2011). Effects of mercuric chloride ($HgCl_2$) on accumulation of total indole alkaloids (vinblastine and vincristine) in calli cultures of Egyptian *C. roseus* were studied (Fathalla et al. 2011). The obtained results showed that the percentage of total indole alkaloids was significantly increased in shoot- and root-derived calli. The highest values of vinblastine and vincristine in shoot- and root-derived calli were recorded at the rate of 0.8 mM $HgCl_2$ in the culture medium.

The presence of cadmium in the plant cell results in the change of enzyme activity and imbalance of micro and macro element distribution and produces lipid peroxidation in the plant cells (Smeets et al. 2008). Cadmium nitrate caused

Table 1.17 Response of different in vitro cultures of *C. roseus* to chemicals

Culture type	Abiotic elicitor	Response	Reference
Hairy root clone LBE-6-1	Nitric oxide (NO) treating the root cultures with 0, 0.1, 1, and 10 mM SNP The samples were harvested on day 9, 14, 17, 20, 23, 26, and 30	0.1 mM SNP treatment on day 0 caused a significant increase in the concentration of serpentine, catharanthine, ajmalicine, lochnericine, and tabersonine production	Li et al. (2011)
		Treatment of 0.1 mM SNP did not affect the growth of hairy roots, while 10 mM SNP almost completely inhibited the growth of hairy roots	
Shoot- and root-derived calli (cv. alba)	$HgCl_2$ was added to culture medium at levels of 0.0, 0.2, 0.4, and 0.8 mM	The highest vincristine content was 128.672 and 117.524 μg/g dw in shoot- and root-derived callus, respectively, grown on B5 medium, whereas the intact plant contained 404.000 μg/g dw vincristine	Fathalla et al. (2011)
Leaf-derived callus (Cultures were maintained at 22 ± 30 °C under a 16 h photoperiod)	Cadmium nitrate $Cd(NO_3)_2$ (10, 30, and 50 mM)	The highest content of total alkaloid was observed with 50 mM $Cd(NO_3)_2$ after 3 days of treatment	Abnosi et al. (2015)
Embryogenic callus from hypocotyl of germinated seedlings and plantlets from somatic embryos	MS + Different levels of NaCl (0.0, 25, 50, 75, 100, and 125 mM)	The addition of NaCl in medium restricted callus growth in almost all tested concentrations An improved synthesis of vinblastine and vincristine was observed by using NaCl (25 mM) as an elicitor	Fatima et al. (2015)
Callus cultures	PEG 4000 at 5%, 10%, and 15%	Maximum vincristine and vinblastine accumulation were 29.650 and 48.000 μg/g dw, respectively, with PEG 15%	Khashan and Al-Athary (2016)

cellular membrane damage and viability reduction but had increased the production of secondary metabolites such as total alkaloids. Therefore cadmium treatment might be a useful way to increase production of alkaloids at cellular level (Abnosi et al. 2015).

Recently, Abnosi et al. (2015) reported that the highest total alkaloid contents of *C. roseus* callus (18.23 mg/g FW) were observed with 50 mM Cd $(NO_3)_2$. Also the activity of superoxide dismutase, catalase, and peroxidase as well as malondialdehyde level as an indicator of lipid peroxidation increased significantly. Cadmium nitrate induced protein profile changes as well as proline content elevation.

Moreover, various in vitro grown embryogenic tissues of *C. roseus* were cultivated under salinity stress for enhanced synthesis of alkaloids. The hypocotyls of germinated seedlings were cultured on MS supplemented with optimized concentration of 2,4-D (1.0 mg/l) for the induction of embryogenic callus. The results revealed that high NaCl concentrations were noted to be inhibitory especially at initiation and proliferation stage of tissues. Quantitative estimation and comparative yield of alkaloids showed that vinblastine and vincristine content was observed to be maximum in regenerated leaves, 14.17 and 5.12 μg/g dry wt, respectively, on medium containing 25 mM NaCl. However, the antioxidant enzyme activity increased linearly with increasing NaCl level in medium (Fatima et al. 2015). More recently, Khashan and Al-Athary (2016) showed that the addition of PEG 15% to the culture medium of *C. roseus* callus tissue resulted in 29.44- and 16.94-fold increase of vinblastine and vincristine, respectively.

4.1.11 Long-Term Preservation

It was suggested that immobilization or entrapment of the cells may produce a microenvironment that resembles the organized tissue in the intact cells causing differentiation and production of secondary metabolites. Few studies have been used for long-term preservation to cell cultures of *C. roseus* to test biosynthetic capacity after immobilization or cryopreservation process. Response of different in vitro cultures of *C. roseus* to long-term preservation is presented in Table 1.18.

Using Ca-alginate beads in immobilization of *C. roseus* cell cultures resulted in a significant increase in the accumulation of ajmalicine in the medium (Asada and Shuler 1989). Also, it was reported that neither growth kinetics nor the degree of vacuolization during growth of *C. roseus* cell cultures was affected by the long-term preservation. The highest content of ajmalicine and catharanthine was 0.24 and 50 mg/g DW in cryopreserved cell cultures. Some changes in secondary metabolism were however found, indicating that preservation under mineral oil does not preserve the productivity of cell cultures whereas the cryogenic method does (Mannonen et al. 1990).

Table 1.18 Response of different in vitro cultures of *C. roseus* to long-term preservation

Culture type	Abiotic elicitor	Response	Reference
Cell suspension culture	Ca-alginate immobilized cells	Immobilization in Ca-alginate beads gave an increase in extracellular ajmalicine of about 3.5 times that for an identical mass of suspended cells	Asada and Shuler (1989)
Cell suspension cultures (Line LD50/h13) originates from anthers of cv. Little Delicata	Preservation in liquid nitrogen or under mineral oil for 6 months	Preservation under mineral oil does not preserve the productivity of cell cultures, whereas the cryogenic method does	Mannonen et al. (1990)

4.1.12 Small- and High-Volume Scale-Up

Stirred-tank bioreactors (STB) equipped with suitable impellers can provide high volumetric mass transfer coefficients and a homogeneous environment enabling suspended plant cell growth production to be controlled consistently (Zhong et al. 2002). Response of different in vitro cultures of *C. roseus* to small- and high-volume scale-up is presented in Table 1.19.

Cell cultures of *C. roseus* were scaled up to volumes of 5000 l using conventional reactors equipped with fiat-blade impellers (Schiel and Berlin 1987). In this study the results showed that catharanthine accumulated only in shake flask experiments. Hoopen et al. (1994) reported that growth of biomass of *C. roseus* cell suspensions was not affected by the type of culture vessel. The production of ajmalicine on production medium in a shake flask was not reproduced in a bioreactor. However, increasing the biomass concentration both in a shake flask and in a stirred fermenter decreased the ajmalicine production rate. Also, Zhen-Gui et al. (1998) found that *C. roseus* cell dry weight and alkaloid production showed no significant difference in the small-volume scale-up shake flask. The maximum cell dry weight was similar between the cell cultures in the shake flask and the bioreactor, but the alkaloid production of cells was much less in the bioreactor. Godoy-Hernandez et al. (2000) found that total alkaloid production in *C. roseus* cell line cultured in a 14 l bioreactor decreased more than 80% while scaling up this cell line from 250 ml batch cultures to the bioreactor.

The important process parameters inside the bioreactor vessel such as pH, temperature, dissolved oxygen, and carbon dioxide evolution rate were investigated (Sivakumar 2006). In this connection, Jeong et al. (2006) stated that alteration of the aeration volumes is the most conventional method of improving the oxygen availability to the growing cultures. In bubble column bioreactors, the volumetric air flow rate is of particular importance as it affects the oxygen transfer, mixing, heat transfer, as well as the degree of turbulence (Srivastava and Srivastava 2012). Cell cultures of *C. roseus* were developed from shake flasks to scale-up through stirred-tank bioreactor (Taha et al. 2014). Results indicated that using uncontrolled pH medium produced 615.35 and 50.76 (g/run) for cells fresh and dry weights, respectively. Moreover, it produced the highest total alkaloid content (1.38% DW), and vincristine and vinblastine content were 5.61- and 14.52-fold, respectively, as compared with intact plant after 15 days of cultivation in stirred reactor. Concerning aeration levels it was found that the capacity of 0.8 l/min was the best as aeration level with uncontrolled pH medium which produced 734.12 and 67.91 (g/run) of cells fresh and dry weights, respectively. However, it produced the highest total alkaloid (1.40% DW). Vincristine and vinblastine content were 13.47- and 7.94-fold, respectively, as compared with intact plant. Recently, Thakore et al. (2017) found that the cultivation of hairy roots in rotating drum bioreactor resulted in ajmalicine content of only 4.6 mg/l. However, the ajmalicine concentration surpassed even that obtained in shake flask (with 34 mg/l; in a customized bioreactor wherein the hairy roots were anchored onto a polyurethane foam, the highest reported in this scale of cultivation).

Table 1.19 Response of different in vitro cultures of *C. roseus* to small- and high-volume scale-up

Culture type	Abiotic elicitor	Response	Reference
Cell suspension cultures (Line CP-3)	Shake flask and fermenter (equipped with flat-bladed impellers)	Catharanthine accumulated only in shake flask Shake flasks yielded ajmalicine (1.3–1.6 mg/g) within 12–15 days	Schiel and Berlin (1987)
Cell suspension cultures	Type of culture vessel (Shake flask and stirred fermenter)	Growth of biomass cell suspensions was not affected by the type of culture vessel Increasing the biomass both in a shake flask and in a stirred fermenter decreased the ajmalicine production	Hoopen et al. (1994)
Cell suspension cultures (Line C20D)	Shake flasks (different volumes, 50, 100, 250, and 500 ml)	No different in the biomass between the cultures grown in different shake flask volumes Maximum alkaloid production was observed on 250 ml shake flask	Zhen-Gui et al. (1998)
	Shake flask (250 ml) and bioreactor (10 l)	The last biomass was equivalent Production of ajmalicine and serpentine was higher in shake flask than bioreactor	
Cell suspension cultures	Cultured in batch culture and then bioreactor	Total alkaloid production decreased more than 80% while scaling up this cell line from 250 ml batch cultures to the bioreactor	Godoy-Hernandez et al. (2000)
Suspension cultures were induced from the friable leaf calli and maintained in an agitated liquid MS medium containing 1.0 mg/l kin and incubated under daylight condition 16/8 at 26 °C and 120 rpm	Controlled or uncontrolled pH and different aeration value rates (0.5, 0.8, and 1.0 l/min)	Capacity of 0.8 l/min was the best as aeration level with uncontrolled pH medium for FW, DW, and total alkaloids, vincristine and vinblastine content	Taha et al. (2014)
Hairy root cultures (Grown on a gyratory shaker rotating at 70 rpm with16/8 h L/D photoperiod at 25 ± 1 °C in shake flask and different bioreactors) (cv. Pra-bal, clone CP 32)	Aeration rates (Different rates (0.3, 0.5, 0.8 vvm)	A maximum of 21 mg/l ajmalicine was observed at the lowest aeration volume (0.3 vvm)	Thakore et al. (2017)

(continued)

Table 1.19 (continued)

Culture type	Abiotic elicitor	Response	Reference
		The large-scale studies show that the ajmalicine production was better in bioreactor with PUF support as compared to shake flask	

4.2 *Biotic Elicitation and TIA Production from In Vitro Cultures of* C. roseus

As the dosage is decisive for the effect, (sterile) biotic elicitor preparations must be standardized, e.g., by the carbohydrate content. The culture age at the moment of addition of the elicitor is another important aspect; often addition a few days after inoculation of the culture, when the cells are rapidly dividing, is preferred. After addition, the elicitor is rapidly dispersed homogeneously throughout the culture medium, and the actual elicitor molecules can rapidly come into contact with their specific receptors in the plant cell wall. Eventually, the signal generated at the cell wall by the elicitor-receptor interaction results in the expression of nuclear genes (Van der Heijden et al. 2004).

4.2.1 Yeast Extract and Chitosan

Yeast extract (YE) is a biotic elicitor; the polysaccharide and the peptide moiety have been recognized as a signaling element in enriching secondary metabolites. YE contains several components like chitin, *N*-acetylglucosamine oligomers, β-glucan, glycopeptides, and ergosterol (Boller 1995), but the specific fraction that triggers the synthesis of alkaloids is still not elucidated clearly. The peptide and polysaccharide moieties present in YE have been suggested to be the molecules of stimulating nature (Menke et al. 1999; Zhao et al. 2011). However, Menke et al. (1999) reported that yeast extract induces transcription of the biosynthetic gene encoding strictosidine synthase (STR) in cultured *C. roseus* cells and alkalinization of the culture medium. Response of in vitro cultures of *C. roseus* to yeast extract and chitosan is presented in Table 1.20.

It is known that the yield of two most investigated alkaloids vinblastine and vincristine is unfortunately very low. A vast array of technologies including elicitation have recently been used to enrich *C. roseus* alkaloid in culture. In this respect, Khashan and Husain (2015) studied effect of yeast extract on vinblastine and vincristine production from leaf-derived callus and leaves of *C. roseus* mother plants. They found that the highest quantity of vinblastine and vincristine (69.56, 35.75 μg/g dw, respectively) was found for yeast extract agar (2 g/l) with significant difference compared with control treatment (1.63 and 1.75 μg/g dw) and mother plant.

Table 1.20 Response of in vitro cultures of *C. roseus* to yeast extract and chitosan

Culture type	Biotic elicitor	Response	Reference
Callus tissues from leaves and leaves of mother plant	Yeast extract agar was added at concentration (1, 2, and 3 g/l)	Addition of 2 g/l yeast extract agar gave 42.67 and 20.42-fold increase for vinblastine and vincristine production, respectively	Khashan and Husain (2015)
Cultivated tissues/ plants developed from protoplast	Yeast extract (0.5, 1.0, 1.5, and 2.0 g/l)	The highest yield of vinblastine (11.44 μg/g dry weight) and vincristine (3.46 μg/g dry weight) was observed in germinating embryos and in in vitro raised leaf 15.47 and 4.14 μg/g dry weight with 1.5 g/l yeast extract	Maqsood and Abdul (2017)
Cell suspension at stationary phase	0, 50, 100, 250, or 500 mg/l medium molecular weight chitosan	100 mg/l medium molecular weight chitosan produces the highest amounts of vinblastine and vincristine	Pliankong et al. (2018)
	Chitosan derived from shrimp shell	100 mg/l chitosan leads to the highest quantity of vinblastine 250 mg/l chitosan leads to the highest quantity of vincristine	

However, the highest value of total soluble carbohydrates (324.444 ppm) and proline (1.208 μmol/g dw) was recorded with yeast extract agar treatment (3.0 g/l). Also, Maqsood and Abdul (2017) studied elicitation in various protoplast-derived tissues and plantlets of *C. roseus* using different concentrations of yeast extract in the culture medium. They found that the highest yield was in medium with 1.5 mg/l yeast extract in which 22.74% vinblastine and 48.49% vincristine enrichment was noted in germinating embryos. Protoplast-derived embryogenic tissue showed higher antioxidant enzyme activities compared to normal embryogenic tissue. It was suggested that the activities of antioxidant enzymes such as ascorbate peroxidase (APX), superoxide dismutase (SOD), and catalase (CAT) are often used as marker in evaluating physiological, biochemical alterations of plants in response to stresses (biotic and abiotic) and in studies with plants' systemic acquired resistance system (Tanabe et al. 2008); and these enzyme activities increase at the onset and increasing levels of stress (Ren et al. 2002).

Recently, Pliankong et al. (2018) showed that after 14 days of culture, the cell suspension of *C. roseus* grown in the 100 mg/l medium molecular weight chitosan produced the highest amounts of vinblastine and vincristine at 4.15 and 5.48 μg/mg cell dry weight, respectively. At the same time, the controls (0.0 mg/l chitosan) produced the two alkaloids at only 2.43 and 2.49 μg/mg cell dry weight, respectively. For chitosan from shrimp shell, it was found that 100 mg/l chitosan leads to the highest quantity of 4.09 μg vinblastine/mg cell dry weight. The highest amount of 5.47 μg vincristine/mg cell dry weight was obtained when 250 mg/l chitosan was added.

4.2.2 Fungal Elicitors

Many secondary metabolites of interest for human health and nutrition are produced by plants when they are under attack of microbial pathogens or insects. Treatment with elicitors derived from phytopathogens can be an effective strategy to increase the yield of specific metabolites obtained from plant cell cultures. Different culture filtrates of *Pythium aphanidermatum*, *Phytophthora cactorum*, *Aspergillum niger*, *Penicillium* sp., and other species were added to cell suspensions and hairy root cultures of *C. roseus* to increase the production of indole alkaloids and phenolics. Response of in vitro cultures of *C. roseus* to fungal elicitors is presented in Table 1.21.

Rapid accumulation of indole alkaloids was observed when cell suspension cultures of *C. roseus* were treated with *Pythium aphanidermatum* (Eilert et al. 1987). Alkaloid formation was preceded by increasing in the activities of the enzymes tryptophan decarboxylase and strictosidine synthase. Asada and Shuler (1989) found that the presence of the fungal elicitor *Phytophthora cactorum* in cell suspension culture of *C. roseus* stimulates a 60% increase in ajmalicine production. Moreno et al. (1993b) reported that elicitation of *C. roseus* cells by *Pythium aphanidermatum* increased tryptamine accumulation in nonfed cells with precursors. A decrease of tryptamine level was also observed, despite the induction of the tryptamine pathway after elicitation. Also, Sim et al. (1994) reported that fungal elicitation by *Penicillium* sp. or *Aspergillus niger* enhanced both alkaloid production and secretion from *C. roseus* hairy root clone LL 15. Alkaloid production reached its maximum when the concentration levels of both fungal elicitors were around 0.01 g/l. However, by combining in situ adsorption by XAD-7 with both fungal elicitation, the release ratio of catharanthine and ajmalicine was enhanced 3.4 and 2 times higher than that obtained with in situ adsorption by XAD-7 alone. Contin et al. (1999) found that elicitation of *C. roseus* cell suspension culture by cellulase and pectinase from *Rhizopus* sp. decreased the cellular contents of secologanin and strictosidine, whereas it increased the tryptamine content. In addition, the uptake of loganin in elicited cells was blocked.

Combined elicitor treatment of an *A. niger* mycelium and tetramethyl ammonium bromide with *C. roseus* cell cultures was used (Zhao et al. 2000). They found that the highest content of ajmalicine was observed in the nutrient medium and in the elicited cell cultures grown in 1000 ml flasks. Abdel-Rahman et al. (2010) demonstrated that *C. roseus* suspension cultures treated with *Alternaria alternata* elicitor, vanadyl sulfate ($VaSO_4$), and potassium chloride (KCl) showed a general increase in alkaloid production and indicated the presence of both extracellular and intracellular vincristine and vinblastine in the cultures. Vincristine quantity was much higher than vinblastine either in extracellular or intracellular samples.

Verma et al. (2014) showed that maximum biomass production of the *Vinca minor* hairy root and cell suspensions was obtained by 10% v/v *T. harzianum*. In case of hairy root, 20% increase in total alkaloid concentration was obtained, and vincamine production was detected. In cell suspensions, maximum of 50% increase

Table 1.21 Response of in vitro cultures of *C. roseus* to fungal elicitors

Culture type	Biotic elicitor	Response	Reference
Cell suspension cultures (line no. 615) from anther explants of greenhouse-grown plants Gyratory shaker (150 rpm) under continuous low light (CV. Little Delicata)	*Pythium aphanidermatum*	Fungal elicitors can be used to facilitate studies of the factors which regulate expression of indole alkaloid pathway enzymes and their ultimate pathway products	Eilert et al. (1987)
Cell suspension line (CP-3) (grown in the dark at 25 °C on a gyratory shaker 120 rpm)	*Phytophthora cactorum* (Elicitor concentrations from 0.5% to 4% (V/V) for exposure time of 48 h)	Stimulates 60% increase in ajmalicine production. The response time to elicitor addition was under 11 h Elicitor concentration (2%) gave the best results	Asada and Shuler (1989)
Cell suspension culture	*Pythium aphanidermatum*	Elicitation increased tryptamine accumulation in nonfed cells, but it did not increase alkaloid accumulation in precursor-fed cells	Moreno et al. (1993b)
Hairy root clones (Cultures were maintained in 250 ml flasks at 25 °C in the dark on a rotary shaker 100 rpm) (cvs. Little Linda- LL and Little Delicata-LD)	*Aspergillus niger* KCTC 1232	When 0.01 g/l *A. niger* homogenate was added, catharanthine production increased 2.5 times	Sim et al. (1994)
	Penicillium sp.	When 0.01 g/l *Penicillium* sp. homogenate was added, both catharanthine and ajmalicine showed a noticeable increase	
Cell suspension cultures (cell line A11)	Cellulase and pectinase (0.002% (w/v) for 15 h) from *Rhizopus* sp.	Elicitation by cellulase and pectinase decreased the cellular contents of secologanin and strictosidine, whereas it increased the tryptamine content	Contin et al. (1999)
Cell suspension culture derived from calli of leaf explants (placed on a rotary shaker (130 rpm) at 23 ± 2 °C in darkness)	*Aspergillus niger* mycelium (20 mg carbohydrate equivalents/l) accompanied with tetramethyl ammonium bromide (120 mg/l)	All elicited cell cultures accumulated ajmalicine between 2.5- and 4-fold increase Serpentine accumulation was about five times higher in elicited cell cultures than that in the control	Zhao et al. (2000)

(continued)

Table 1.21 (continued)

Culture type	Biotic elicitor	Response	Reference
		Elicited cell cultures also accumulated catharanthine by about three times than the control	
		Elicited cell cultures had lower biomass yield than the control Lower biomass yield was obtained in the bioreactor than in 500 and 1000 ml flasks	
		Increase of malondialdehyde content in cell cultures	
Cell suspensions (Cultures were maintained in gyratory shaker at 150 rpm at 27 °C in dark)	*Alternaria alternata*, vanadyl sulfate ($VaSO_4$), and potassium chloride (KCl)	HPLC quantitative estimations of alkaloids showed that extracellular alkaloids were much more than the intracellular ones	Abdel-Rahman et al. (2010)
Hairy roots and cell suspensions *(V. minor)*	*Chaetomium globosum; Aspergillus niveoglaucus; Paecilomyces lilacinus; Trichoderma harzianum* (Culture filtrates at the dose of 3.3%, 6.0%, and 10% v/v.)	Culture filtrate of *T. harzianum* (10% v/v) positively enhanced the biomass accumulation and alkaloid content of hairy roots and cell suspensions both in the shake flask culture and the bioreactor	Verma et al. (2014)
Callus tissues from leaves of mother plant	*Trichoderma harzianum* and *Azotobacter chroococcum* bacteria	Treatment with *T. harzianum* gave 36.66- and 19.94-fold increase for vinblastine and vincristine production, respectively	Khashan and Husain (2015)
Embryogenic callus Germinating embryos	*Aspergillus flavus* (0.0%, 0.05%, 0.15%, 0.25%, and 0.35%)	*A. flavus* elicitation at 0.15% improved vinblastine yield	Tonk et al. (2016)
Maturated and germinating somatic embryos		Yielded enriched level of vinblastine and vincristine at 0.05% and 0.15% levels of *A. flavus*	
Suspensions of cambial meristematic cells (CMCs) grown at 25 °C at 110 rpm with a 24 h light photoperiod. After culturing for another 12, 24, 36, 48, and 60 h, the cultures and liquid media were harvested	*Aspergillus flavus* mycelium elicitor or *A. flavus* medium elicitor at final concentrations of 5, 15, and 25 mg/l	Addition of *A. flavus* (25 mg/l) mycelium elicitor to suspension culture after 6 days of cultivation resulted on increasing the contents of vindoline, catharanthine, and ajmalicine	Liang et al. (2018)

in the total alkaloid content was observed in the 10% v/v *T. harzianum* application. None of the cell suspension sample showed presence of vincamine. Treated hairy root clone PVG and cell suspension were upscaled in 5 l stirred-tank bioreactor with the addition of 10% v/v *T. harzianum* culture filtrate. The results showed mark improvement in biomass accumulation and total alkaloid in the third run and the six run for hairy roots and cell suspension. Moreover, the shake flask-grown hairy roots showed 4.5-fold increase in the tryptophan decarboxylase (TDC) expression, while the bioreactor harvested hairy roots showed 6-fold enhancement. Similarly, cell suspension showed 2- and 2.5-fold enhanced TDC expression in shake flask and bioreactor, respectively. In case of strictosidine synthase (STR), 5.5- and 6.5-fold increase in the expression level was registered by the hairy roots from the shake flask and the bioreactor, respectively. The cell suspensions also showed more than threefold enhanced STR expression level both in the shake flask and the bioreactor.

The effect of fungi (*Trichoderma harzianum*) and bacteria (*Azotobacter chroococcum*) on vinblastine and vincristine production from *C. roseus* callus tissue and leaves of mother plant was investigated (Khashan and Husain 2015). They found that the highest quantity of vinblastine (59.760 μg/g dw) and vincristine (34.900 μg/g dw) was recorded with the fungus *T. harzianum* with significant difference compared with control treatment and mother plant. Moreover, the highest catalase enzyme activity (61.49 unit), total soluble carbohydrates (315.330 ppm), and proline (0.928 μmol/g dw) were recorded with *Trichoderma harzianum* treatment and were more than control and leaves of intact plant. Recently, Tonk et al. (2016) reported that the use of *Aspergillus flavus* fungal elicitation improved callus biomass growth in callus tissues of *C. roseus*. *A. flavus* elicitation at 0.15% improved vinblastine yield (0.903 μg/gm dry weight). However, the yield of vincristine was low, and on *A. flavus* addition, maximum vincristine yield was noted, 0.216 μg/g dry weight. Maturated and germinating somatic embryos yielded enriched level of vinblastine and vincristine at 0.05% and 0.15% levels of *A. flavus*. More recently, Liang et al. (2018) found that addition of *A. flavus* (25 mg/l) mycelium elicitor to suspension culture of *C. roseus* cambial meristematic cells (CMCs) after 6 days of cultivation and the CMC suspensions were further cultured for another 48 h and resulted in increasing the contents of vindoline, catharanthine, and ajmalicine, 1.45, 3.29, and 2.14 times as high as those of the control group, respectively. Transcriptome analysis showed that D4H, G10H, GES, IRS, LAMT, SGD, STR, TDC, and ORCA3 were involved in the regulation of this induction process. The results of qRT-PCR indicated that the increasing accumulations of vindoline, catharanthine, and ajmalicine in CMCs were correlated with the increasing expression of the above genes.

4.3 *Precursor Feeding and TIA Production from In Vitro Cultures of* C. roseus

There are various ways to increase secondary metabolite content in tissue culture; one of them is precursor feeding. Feeding with specific biosynthetic precursors or metabolites either individual or combined has proved to be a successful strategy to increase the biosynthesis of indole alkaloids in *C. roseus* cell and hairy root cultures. Studies on *C. roseus*, specifically, tryptophan, tryptamine, loganin, secologanin, catharanthine, stemmadenine, and geraniol 10-hydroxygeraniol have been act as precursors most used. Attempts to increase the yield of alkaloids in cell and transgenic cultures of *C. roseus* by supplying precursors or intermediate compounds are found to be effective in many cases. However, in order to evaluate the flux-limiting steps of the TIA pathway, precursor feeding studies were performed in the *C. roseus* cell cultures. Response of in vitro cultures of *C. roseus* to precursor feeding is presented in Table 1.22.

Addition of tryptophan to suspension cultures of *C. roseus* did not affect production of ajmalicine and serpentine, whereas addition of secologanin enhanced alkaloid production (Mérillon et al. 1986). Also, addition of secologanin, loganin, or loganic acid to suspension cultures of *C. roseus* caused an increase in the accumulation of ajmalicine and strictosidine. In precursor-fed cells, there is no further increase in alkaloid accumulation despite of exposure cells to elicitation by *Pythium aphanidermatum*. Feeding mevalonic acid did not increase alkaloid accumulation in any studied case (Moreno et al. 1993b). It was found that, when cells were grown in 500 ml medium supplemented with tryptophan (250 mg/l), the growth pattern of the cells was similar to growth medium. Maximum production of ajmalicine was 315 μg/g dry weight; after 14 days of cultivation was observed in single-stage bioreactor, while double-stage bioreactor yielded trace amount of ajmalicine (Fulzele and Heble 1994). Extensive precursor feeding experiments were performed using transgenic *C. roseus* hairy roots or cell lines. Addition of 500 μM loganin to the transgenic cell cultures gave 263.6 μmol/l total alkaloid, while the control treatment gave 123.7 μmol/l. However, alkaloid content of transgenic cell culture treated with loganin and tryptamine was 350 μmol/l, while the control gave 198 μmol/l (Whitmer et al. 1998).

Cultures of transgenic *C. roseus* consistently showed tenfold higher STR activity than wild-type cultures. Addition of loganin to the medium greatly enhanced alkaloid production, reaching up to 280 mg/l, including ajmalicine, catharanthine, serpentine, and tabersonine, while maintaining wild-type levels of TDC activity. Alkaloid accumulation by highly productive transgenic lines showed considerable instability and was strongly influenced by culture conditions and the availability of precursors (Canel et al. 1998). In another study, saturation in the uptake of loganin and in the conversion of loganin into secologanin in *C. roseus* cell culture was observed after feeding with increasing amounts of loganin. However, feeding of tryptophan did not result in any increase in alkaloid contents, despite its complete uptake. Also, tryptamine feeding led to increased strictosidine contents, but ajmalicine levels remained unchanged (Contin et al. 1999). In another work, Morgan and Shanks (2000) found that addition of low levels of auxin or tryptophan to *C. roseus* hairy root cultures resulted in significant increases in flux to the indole alkaloids. However, the addition of tabersonine (40 mg/l) to cultures at day 21 resulted in the increased yield of tabersonine and the decreased yield of horhammericine. Feeding of geraniol, 10-hydroxygeraniol, or loganin at 21 days each resulted in significant increases in the accumulation of tabersonine. Feeding compact callus cluster from *C. roseus* with succinic acid, tryptamine, and tryptophan significantly increased

Table 1.22 Response of in vitro cultures of *C. roseus* to precursor feeding

Culture type	Precursor	Response	Reference
Cell suspension culture (The cells were agitated on a gyratory shaker, at 24 °C, in the dark)	Tryptophan (Trp) (1.22 mM) or/ and secologanin (Sec) (0.124 mM)	Addition of secologanin enhanced ajmalicine and serpentine production whatever the culture medium used	Mérillon et al. (1986)
Cell suspension culture	Secologanin, loganin, or loganic acid	Increase in the accumulation of ajmalicine and strictosidine and a decrease of tryptamine level were observed in non-elicited cells	Moreno et al. (1993b)
	Mevalonic acid	Did not increase alkaloid accumulation in any studied case	
Cell suspension cultures (20-1 airlift bioreactor) Single-stage bioreactor	Tryptophan (250 mg/l)	The higher level of ajmalicine was observed in production medium enriched with 250 mg/l tryptophan	Fulzele and Heble (1994)
Transgenic cell suspension cultures (Line S10 carries a recombinant constitutively overexpressed version of the endogenous strictosidine synthase (Str) gene (var. Morning Mist))	Tryptamine	Tryptamine enhances metabolic flux through the indole pathway	Whitmer et al. (1998)
	Loganin (500 μM)	Increased total alkaloid than control treatment and decreased tryptamine content than control	
	Addition of loganin + tryptamine	Increased total alkaloid than control treatment	
Transgenic cell suspension cultures (overexpress the enzymes strictosidine synthase (STR) and tryptophan decarboxylase (TDC))	Loganin (100 mM)	Alkaloid accumulation by highly productive transgenic lines showed considerable instability and was strongly influenced by the availability of precursors	Canel et al. (1998)
Cell suspension cultures (cell line A11)	Loganin (500 μM)	Four-day-old cells converted 100% of the fed loganin to secologanin within 24 h	Contin et al. (1999)
	Loganin (10–10,000 μM)	Increasing amounts of loganin result in uptake of loganin and conversion of loganin into secologanin	

	Tryptophan and tryptamine	Feeding of tryptophan did not result in any increase in alkaloid contents Tryptamine feeding led to increased strictosidine contents	
Hairy root cultures LBE 6-1 (Cultures were shaken at 100 rpm and incubated at 26 °C in the dark)	Tryptophan and auxin (NAA or IAA)	Increased branching and thickening of the hairy root cultures. A dramatic reduction in flux to the alkaloids was observed	Morgan and Shanks (2000)
	Tryptamine or terpenoid precursors	Had no effect	
	Addition of tryptophan or tryptamine	Had no effect during the stationary phase of the growth cycle	
	Geraniol, 10-hydroxygeraniol, or loganin at 21 days	Resulted in significant increases in the accumulation of tabersonine	
	Combined elicitation with jasmonic acid (50 mg/l) and feeding either loganin (40 mg/l) or tryptamine (20 mg/l)	Did not further enhance the accumulation of indole alkaloids	
Compact callus cluster	Succinic acid (10 mM)	Resulted in the highest ajmalicine and serpentine levels The levels of which were about five- and fourfold higher than in the control	Zhao et al. (2001a)
	Tryptamine (300 and 500 mg/l)	Resulted in the highest amounts of catharanthine and 500 mg/l tryptamine gave the highest yield of ajmalicine	
	Tryptophan (300 and 500 mg/l)	Showed less improving effects	
	Geraniol feeding	Inhibited biomass and alkaloid accumulation	
Cell suspension culture	Loganin and tryptamine	Formation of strictosidine which is further converted into ajmalicine	El-Sayed and Verpoorte (2002)
Transgenic cell line (line T22) over expressing tryptophan decarboxylase (Tdc)	Tryptophan, tryptamine, loganin, and secologanin	Addition of loganin or secologanin enhanced alkaloid production	Whitmer et al. (2002)

(continued)

Table 1.22 (continued)

Culture type	Precursor	Response	Reference
Cell suspension culture (Cultures were shaken at 110 ± 5 rpm and at 25 ± 1 °C, with 24 h light)	Stemmadenine (mg per flask)	Accumulation of catharanthine, tabersonine, and condylocarpine	El-Sayed et al. (2004)
Hairy root line expressing the anthranilate synthase α subunit	Feeding with 1-deoxy-D-xylulose (208 mM)	Increase of 125% in horhammericine levels	Peebles et al. (2006)
	Feeding with loganin	Increased catharanthine by 45%	
	Feeding with loganin (104 mM)	Increases catharanthine by 26%, ajmalicine by 84%, lochnericine by 119%, and tabersonine by 225% in the induced samples	
Hairy root cultures (var. Nirmal)	Loganin	Ajmalicine production was improved by 2.3-fold, and there was 1.8-fold increase in the yield of serpentine when compared with control cultures	Gaviraj and Veeresham (2006)
	Multiple feedings of loganin	Twofold improvements for the production of ajmalicine and serpentine over the control cultures Catharanthine level was improved by 1.9-fold over the control	
	Loganin + tryptamine	Catharanthine and ajmalicine accumulation was not affected Serpentine production was 2.4-fold increase over the control	
Cell suspension cultures (line C20D) on a rotary shaker (100 rpm) at 24 °C in the dark	Loganin	Addition of the loganin in the fourth day, to cell culture medium subjected to hypoxia restored the alkaloid production	Senoussi et al. (2007)
Leaf discs	Phenylalanine (50, 100, and 200 ppm) Tryptophan (50, 100, 200, 400, and 600 ppm)	The precursor phenylalanine (50 ppm) showed a better result than tryptophan	Al Khateeb et al. (2009)
EHIORCA3-4-1 hairy roots (Hairy roots containing an inducible ORCA3 gene)	Loganin (104 μM) and/or tryptophan (104 μM)	The highest content of TIAs was recorded with hairy roots containing an inducible ORCA3 gene which fed an equal volume of ethanol	Peebles et al. (2009)

Noninduced and MeJA-induced hairy root cultures	Loganin (0.4 mM) and tryptamine (0.2 mM)	Addition of loganin, tryptamine, or their combination did not increase strictosidine, ajmalicine, serpentine, and tabersonine production in noninduced cultures	Goklany et al. (2009)
		In MeJA-induced cultures, loganin or tryptamine feeding also did not enhance strictosidine, ajmalicine, serpentine, and tabersonine production	
		Secologanin levels in MeJA-elicited cultures fed with loganin, tryptamine, or their combination were 110%, 135%, and 220% increase in levels when compared with nonfed MeJA-elicited cultures	
Suspension cultures derived from leaf calli cultures and incubated under daylight condition 16/8 at 114 rpm	L-tryptophan; L-glutamine; L-asparagine; L-cysteine; and L-arginine at the concentrations of 0; 100; 300; or 500 mg/l	300 mg/l of L-glutamine gave the highest callus biomass and 300 mg/l L-tryptophan enhanced total alkaloid content	Taha et al. (2009)
Cell aggregate culture in airlift bioreactor 1.5 l volume aeration was 0.15 l/min with drought tube	Addition of tryptophan (50–250 mg/l)	The addition of tryptophan induced protein synthesis and TDC activity in aggregate cell culture	Pandiangan et al. (2010)
Cell aggregate cultures incubated at room temperature and agitated at a speed of 120 rpm	Addition of tryptophan (50–250 mg/l)	The highest content of catharanthine occurred on day 14 after 150 mg/l treatment of tryptophan	Pandiangan et al. (2013)
Callus cultures derived from leaf explants Cell suspension cultures were shaken at 120 rpm and incubated at room temperature	Addition of tryptophan (50–250 mg/l)	Tryptamine content in cell suspension increased with increasing levels of tryptophan Catharanthine content in cell suspension increased and has optimum content with tryptophan treatment (150 mg/l)	Pandiangan et al. (2015)
		After day 21 of culture, vinblastine content is detected and has increased as a result of tryptophan treatment	

(continued)

Table 1.22 (continued)

Culture type	Precursor	Response	Reference
Cell suspension culture derived from calli of leaf explants (suspension cultured cells were grown at 25 °C with shaking (110 rpm) for 12 days with 24 h light)	Catharanthine (Biotransformation)	The dry cell weight progressively decreased with the increasing dosage of catharanthine The highest content of total new product (3-hydroxy-4-imino-catharanthine) occurred at 120 h: 0.61 mg/100 ml culture, and the conversion was 9.75%	He et al. (2015)
Callus cultures derived from leaf explants	Tryptophan treatments with concentrations (0.0, 50, 100, 150, 200, and 250 mg/l)	The treatment of 250 mg/l tryptophan gave the highest ajmalicine content after control treatment (0.0 mg/l tryptophan)	Mandagi et al. (2017)
Shoot cultures (maintained at 25 ± 2 °C, under 16 h photoperiod by cool daylight)	Tryptamine (300, 500, and 1000 mg/l) L-tryptophan (300, 500, and 1000 mg/l)	Shoot cultures treated with tryptamine (300 mg/l) accumulated the highest vinblastine content (0.0277% dry wt) followed by tryptophan feeding at 300 mg/l (0.0180% dry wt) and 500 mg/l (0.0175% dry wt), whereas callus cultures failed to produce vinblastine	Sharma et al. (2019)
Callus cultures derived from leaf explants		Callus cultures treated with tryptophan and tryptamine showed a variable decrease in vindoline production compared to non-treated control In case of catharanthine production, callus cultures fed with 500 mg/l tryptophan recorded the trivial increase of 5% with a decrease in cell growth compared to non-treated control	

ajmalicine (41.5, 36.9, and 31.8 mg/l, respectively) and catharanthine (21.1, 17.2, and 18 mg/l respectively; production while geraniol feeding inhibited biomass and alkaloid accumulation (Zhao et al. 2001a). However, El-Sayed and Verpoorte (2002) found that feeding *C. roseus* cell suspension cultures with tryptamine and loganin resulted in high accumulation of strictosidine, and an increase of sixfold in ajmalicine was detected.

Alkaloid accumulation by transgenic cell line overexpressing tryptophan decarboxylase (Tdc), thus having a high level of tryptamine (line T22), was enhanced by addition of loganin or secologanin (Whitmer et al. 2002). They found that feeding tryptamine or tryptophan alone had no effect on TIA accumulation. However, in combination with tryptophan or tryptamine, feeding of these precursors resulted in an even further increase of alkaloid accumulation, and under optimal conditions, line T22 accumulated around 1200 μmol/l of TIAs, whereas the control cultures accumulated less than 10 μmol/l TIAs. In another study, it was found that feeding stemmadenine to *C. roseus* cell suspension culture resulted in the accumulation of catharanthine, tabersonine, and condylocarpine (El-Sayed et al. 2004).

Metabolic engineering has focused on increasing flux through this pathway by various means such as introduction of genes encoding specific metabolic enzymes into the plant. In this respect, Pasquali et al. (1992) stated that the genes Tdc and Str are transcriptionally regulated in *C. roseus*. These genes, respectively, encode the enzyme tryptophan decarboxylase (TDC), an important link between primary and secondary metabolism catalyzing the conversion of tryptophan into tryptamine and strictosidine synthase (STR), which condenses tryptamine and the iridoid secologanin to yield strictosidine, the universal precursor of TIAs. Peebles et al. (2006) reported that by feeding 1-deoxy-D-xylulose to the hairy root line expressing the anthranilate synthase ά subunit, an increase of 125% in horhammericine levels was recorded in the induced samples, while loganin feeding increased catharanthine by 45%. Loganin feeding to the hairy root line expressing anthranilate synthase ά and β subunits increases catharanthine by 26%, ajmalicine by 84%, lochnericine by 119%, and tabersonine by 225% in the induced samples.

Gaviraj and Veeresham (2006) reported that ajmalicine production of *C. roseus* transformed root cultures was improved by 2.3-fold and there was 1.8-fold increase in the yield of serpentine compared with control cultures, when loganin was added to the culture medium. There was 2.4-fold improvement in serpentine production (3.27 mg/l) over the control. Moreover, multiple feedings of loganin increased the accumulation of alkaloids. In addition, there are twofold improvements for the production of ajmalicine (2.77 mg/l) and serpentine (3.04 mg/l) over the control cultures (1.42 and 1.42 mg/l). Catharanthine level (2.53 mg/l) was improved by 1.9-fold over the control (1.31 mg/l).

The relationship between hypoxia and alkaloid production in cell culture of *C. roseus* was investigated (Senoussi et al. 2007). In this study, the addition of the loganin in the fourth day, in the cell culture medium subjected to hypoxia, restored the alkaloid production. Al Khateeb et al. (2009) showed that the precursor phenylalanine (50 PPM) gave the highest alkaloid, in leaf disc cultures of *C. roseus*, while tryptophan showed negative response. However, Peebles et al. (2009) found that *C. roseus* hairy roots containing an inducible ORCA3 gene which fed with loganin and/or tryptophan recorded the highest content (4996 μg/g DW) of TIAs with hairy roots containing an inducible ORCA3 gene which fed an equal volume of ethanol. Also, Goklany et al. (2009) found that TIA production of *C. roseus* hairy root cultures was not significantly enhanced in either noninduced or MeJA-induced cultures with precursor feeding. However, secologanin levels in MeJA-elicited cultures fed with loganin, tryptamine, or their combination were 100.1, 112.0, and 151.1 mg/l when compared with nonfed MeJA-elicited cultures. Secologanin accumulated in MeJA-induced cultures with precursor feeding, suggesting that STR was likely limiting under MeJA-induced conditions. Taha et al. (2009) showed that the highest value of mass cell cultures of *C. roseus* was obtained with addition of 300 mg/l L-glutamine to MS medium, while addition of 300 mg/l L-tryptophan enhanced total indole alkaloids, vinblastine and vincristine as compared with other used amino acids. Pandiangan et al. (2010) reported that the maximum growth of cell aggregate culture of *C. roseus* occurred in the treatment of tryptophan 150 mg/l in airlift bioreactor after 10 days of culture. The highest protein content (2.07 mg/ml) and TDC activity (3.459 units) occurred in the treatment of 250 mg/l tryptophan after 10 days of culture. A study on changes in cell morphology and its association with increased catharanthine content in *C. roseus* cell cultures after tryptophan treatment has been carried out (Pandiangan et al. 2013). The results showed that cell morphology was essentially the same for each treatment. The apparent discrepancy was the presence of long cells in higher number in the treatment of tryptophan. There was relation between changes in cell morphology (more long cells) and the increased content of catharanthine after tryptophan treatment. The highest content of catharanthine (50.96 μg/g dw) occurred on day 14 after 150 mg/l treatment of tryptophan with 75% long cells (specialized cells).

However, Pandiangan et al. (2015) found that tryptamine content in *C. roseus* cell culture increased with increasing concentrations of tryptophan treatment. Vindoline content in cells after 14 days cultivation decreased by treatment with 100–200 mg/l tryptophan. Catharanthine content of cells was increased with tryptophan (150 mg/l) that was equal to 50.96 μg/g DW. However, the content of vinblastine in cells until day 14 culture is not detected, but after day 21 vinblastine content is detected and has increased according to application of tryptophan precursor treatment.

Biotransformation of catharanthine by addition of catharanthine to suspension-cultured cells of *C. roseus* gave a new oxidative cleavage product. The structure of this product was determined as 3-hydroxy-4-imino-catharanthine by spectroscopic methods (He et al. 2015). After an additional 24 h, the content of product in extracellular medium was higher than that in cells. The result can therefore be explained by the continuous transportation of product through the plasma membrane into the medium. Recently, Mandagi et al. (2017) reported that the highest dry weight (0.26 g) of *C. roseus* callus was at the treatment of 50 mg/l tryptophan. The highest ajmalicine content (9.048 μg/g dw) was detected in control treatment followed by 250 mg/l tryptophan treatment which produced 2.68 μg/g dw, while the lowest ajmalicine content (0.346 μg/g dw) was observed with tryptophan treatment at

200 mg/l. It can be concluded that addition of the tryptophan in callus cultures can decrease the content of ajmalicine. More recently, Sharma et al. (2019) reported that highest alkaloid levels of 6.2% dry weight were found in shoot cultures of *C. roseus* that were fed with 1000 mg/l tryptamine. Tryptamine at its low concentration of 300 mg/l also showed an increase in catharanthine production but no concurrent influences on growth. Tryptophan was also found favoring vinblastine accumulation in cultured shoots. The effect was more prominent (>fivefold) at lower tryptophan supplementation doses of 300–500 mg/l. With regard to callus cultures, tryptamine and tryptophan treatments also failed to enhance biomass productivity to any significant level when compared with non-treated controls. Increase in tryptamine doses at 300, 500, and 1000 mg/l doses caused a decrease in alkaloid content with 1.98%, 1.68%, and 1.55% dry wt. A slight increase in expression of *CrSGD* gene in the tissues treated with tryptophan (300 and 500 mg/l).

5 Conclusion

A survey of literature on in vitro culture of *C. roseus*, including callus, suspension, and transgenic hairy root cultures, is reported in this chapter. A suitable and update example has been provided to the different approaches utilized for increasing the production of alkaloids in this important medicinal plant. In order to establish the different in vitro cultures, researchers used various explants from germinated seeds in vitro as well as meristematic parts from grown field plants. The alteration of culture media, its components, and manipulation of plant growth regulators are also often essential to restore growth of callus cultures. The callus tissue, which may be a source of alkaloids itself, most commonly provides a starting material for establishing the suspension culture or for plant organ induction, i.e., shoots or roots.

Preservation of cell cultures of *C. roseus* at ultralow temperatures has gained increasing interest, and some successful reports on the use of cryogenic storage of cell cultures have been published. Many authors confirmed the feasibility of storing *C. roseus* cells cultured in vitro at the temperature of liquid nitrogen. A minimal amount of liquid water is essential for a cell to survive freezing; it appears that the amount of liquid water at −40 °C is critical for a successful cryopreservation. However, in most studies, the cryopreserved cells retained their alkaloid-producing capability. In general, considerable experimentation is always needed to optimize all the steps during the cryopreservation procedure.

In order to improve the production of TIAs in *C. roseus*, the establishment of an efficient genetic transformation method is required. The transient transformation efficiency can be improved with the sonication treatment. However, there are only few reports pertaining to development of transgenic *C. roseus* overexpressing TIA pathway genes and enzymes, which involved in the biosynthesis of alkaloids. In recent studies tryptophan decarboxylase (TDC) and strictosidine synthase (STR) genes from *C. roseus* have been successfully overexpressed in the *rol* gene integrated cell suspensions of *V. minor*. The physiological state of tissue to be transformed is very crucial for overexpressing the terpenoid indole alkaloid pathway genes. The Ri-T-DNA-incorporated cell suspensions can be utilized as an ideal material for TDC + STR overexpression studies due to the existence of auxin-synthesizing genes in the chromosomes. Hairy roots of *C. roseus* have been successfully established. Survey of literature confirmed the presence of high levels of monoterpenoid indole alkaloids, mainly catharanthine, serpentine, and ajmalicine, in hairy roots, while the anticancer, like vinblastine and vincristine, is absent due to lack of vindoline associated with chloroplasts which is absent in root cells. In contrast, accumulation of vinblastine, vincristine, and catharanthine in the transgenic hairy root was reported.

With regard to elicitation of *C. roseus* cultures, both abiotic, biotic elicitors and precursor feeding have successfully been applied to increase alkaloid production. For the characterization of growth and alkaloid production of *C. roseus* cell cultures, detailed protocols are available. Using abiotic elicitors with in vitro cultures of *C. roseus* showed that the growth with the optimized nutrient concentrations in the medium was more than 50% better than in the control medium with about the same alkaloid production. However, the alkaloid qualitative profile changed depending on the nitrogen source, indicating that alteration of the primary N metabolism does not affect the total amount of nitrogen and carbon funneled into alkaloid metabolism but rather on its distribution. It was also reported that the BAP can without doubt decrease the effects of the hypoxia and increase the ajmalicine production. Moreover, alkaloid levels were also found to be sensitive to the concentration of sucrose in the medium. It was found that the addition of sucrose 60 g/l to the culture medium of *C. roseus* callus tissue resulted in 19.83- and 9.28-fold increase for vinblastine and vincristine, respectively, while addition of PEG (15%) to the culture medium resulted in 29.44- and 16.94-fold increase for vinblastine and vincristine, respectively. Fructose also enhanced the catharanthine yield about twofold in hairy root cultures of *C. roseus*. It was found that buffers had slightly negative effects upon growth of *C. roseus* hairy root; some component (lochnericine) was significantly lower in response to the addition of buffers, while another component (tabersonine) was significantly higher. It was shown that the subculture cycle affected both growth and alkaloid production. Increase in biomass concentration (high-density cultures) did not enhance the total alkaloid production. The major differences are observed in enzyme levels between high- and low-density cultures. Low-density culture had accumulated six times more ajmalicine than the high-density culture.

Concerning the effect of growth regulators, it was reported that the presence of NAA and Kin in the culture medium of *C. roseus* resulted in the highest vindoline, catharanthine, and vincristine contents. Also, methyl jasmonate induced strictosidine and ajmalicine accumulation. Shoot cultures of *C. roseus* subjected to methyl jasmonate also produced up to an eightfold increase of both deacetoxyvindoline 4-hydroxylase (D4H) and deacetylvindoline acetyl CoA acetyltransferase (DAT) activities. It is possible to suggest that MeJA could be regulating the expression of some of the transporters involved in the secretion of the compounds produced by the *C. roseus* hairy roots.

Studies of light showed that under continuous light conditions, the total amount of TIAs in suspension culture of *C. roseus* was increased about 2.8-fold compared to that in the control. Moreover, effects of light treatment on vindoline synthesis were not mediated by ORCA-3 proteins (which are involved in the induction of alkaloid synthesis in response to elicitation), suggesting the presence of a different set of regulatory proteins. However, exposure of *C. roseus* cell suspension culture to low dose of UV-B irradiation was found to increase the amount of catharanthine and transcription of genes encoding tryptophan decarboxylase (*Tdc*) and strictosidine synthase (*Str*). It would be indicated that the molecular basis of UV-B signaling cascades leading to the induction of expression of *Tdc* and *Str* genes and the production of TIAs is largely unknown.

Various workers showed that lowering temperature caused an increase in alkaloid accumulation. It was revealed that the culture conditions promoted gene expression of enzymes in the TIA pathway in *C. roseus* cultures. Studies on hypoxia of *C. roseus* cultures showed that the lack of oxygenation in *C. roseus* cells provokes a very strong inhibition in accumulation of the alkaloids and of other possible substances. This can be explained by the absence of some downstream enzymes involved in the biosynthesis chain of alkaloids or by the blockage of the upstream terpenes to the production stage.

Bioregulators have been found to enhance the flux into alkaloid pathway, thereby improving the yields. Both the P450 inducer phenobarbital and the phenylalanine ammonia-lyase inhibitor trans-cinnamic acid effectively improved serpentine and ajmalicine production in compact cell cultures. Cytochrome P450 enzymes are widely involved in the biosynthesis of indole alkaloids in *C. roseus*, and trans-cinnamic acid blocks the shikimic acid pathway flowing to phenylpropanoid and therefore has been proposed to enhance tryptamine pools for indole alkaloid biosynthesis. Concerning the permeabilizing agents, it was reported that both DMSO and Triton X-100 were found to be useful in effluxing ajmalicine from *C. roseus* hairy root cultures into the medium resulting in a 12- and 16-fold increase, respectively. Addition of $HgCl_2$ or cadmium nitrate Cd $(NO_3)_2$ increased significantly the content of total indole alkaloids in shoot- and root-derived calli of *C. roseus*.

Using biotic elicitors with in vitro cultures of *C. roseus* showed that yeast extract and chitosan significantly affect the accumulation of TIA in different cultures of *C. roseus*. Also, the presence of a fungal elicitor may accelerate the physiological/molecular process leading to the release of alkaloids in a *C. roseus* cell culture. Moreover, the effect of feeding precursors on growth of *C. roseus* cultures varies depending on the cell line and the type of precursor used. Studies on precursor feeding to different cultures of *C. roseus* showed that the utilization of tryptamine (as a precursor) for alkaloid biosynthesis enhances metabolic flux through the indole pathway. Feeding *C. roseus* cell suspension cultures with tryptamine and loganin resulted in high accumulation of strictosidine, and an increase of sixfold in ajmalicine was detected. Generally, these results suggest that the terpenoid precursors are important factors to the terpenoid indole alkaloid production. However, some studies found that feeding tryptophan or tryptamine in *C. roseus* cell cultures increased, decreased, or had no effect on TIA levels.

6 Future Aspects

Studies on production of alkaloid from *C. roseus* cell and transgenic hairy root culture have begun to form a picture of the complex network that leads to the formation of the terpenoid indole alkaloids. Even though there are still many unknown compounds and uncharacterized enzymes within this pathway (Hughes and Shanks 2002; Van der Heijden et al. 2004), researchers can focus on metabolic engineering of the known enzymes and compounds in order to further characterize this complex network. The relationship between UV-B signaling and other TIA signaling pathways is yet to be examined. However, the effect of the permeabilizing agents is species specific, and no generalized theories are therefore available to interpret their mechanism of action. Therefore it becomes important to select the right permeabilizing agent and optimize its dosage and time for better response from the system. In addition more efficient bioreactors for the production and recovery of secondary metabolites from plant cell cultures of *C. roseus* are needed. Precursor feeding studies in particular can help identify limiting enzymes in the pathway in order to focus future research efforts on these limiting enzymes (Peebles et al. 2006). More studies on the signal transduction pathways that link elicitor perception to the biosynthesis of specific secondary metabolites and high-throughput analysis of gene transcripts, proteins, and metabolites in elicited plants will be essential for better understanding the regulation of plant secondary metabolism and for engineering new plants with improved production of metabolic of interest.

Advances in the understanding of the TIA biosynthetic pathway at the DNA level and the possibility of efficiently transferring genes into plant cells offer novel opportunities to improve alkaloid yields by cell cultures. However, synthetic biology which can be defined as "the design and construction of new biological components, such as enzymes, genetic circuits, and cells, or the redesign of existing biological systems" (Keasling 2008) could open new approaches for TIA biotechnological production, based on the transfer of biosynthetic pathways, totally or partially, to faster-growing organisms like yeast or other plant cell cultures. Current progress in the application of synthetic biology technologies has allowed the production of high added value secondary compounds in heterologous systems to become almost a reality. Extensive research is needed to learn more on the regulation of the TIA biosynthetic pathway in *C. roseus* to apply this knowledge to increase the flux toward the compounds of interest. The definitive breakthrough in a cost-effective and sustainable commercial production of plant secondary metabolites will depend on a more in-depth knowledge of the metabolic response to elicitation in plant cells, as well as a thorough understanding of the mechanisms responsible for these effects. Finally, despite the large scientific achievements on in vitro culture of *C. roseus* plant, there is no commercial biotechnological production of these alkaloids yet.

References

Abdel-Rahman TM, Kapiel TYS, Ibrahiem DM, Ali EAM (2010) Elicitation of alkaloids by biotic and abiotic stress factors in *Catharanthus roseus*. Egypt J Bot Special Volume:207–224

Abnosi MH, Amirjani M, Mahdiyeh M, Moradipoor H (2015) Biochemical and cellular response of *Catharanthus roseus* callus cells to cadmium toxicity. J Genet Resour 1(2):101–114

Abou Dahab TAM, Abd El-Aziz GN (2006) Physiological effect of diphenylamine and tryptophan on the growth and chemical constituents of *Phylodendron erubescens* plants. World J Agric Sci 2:75–81

Al Khateeb E, Alani NK, Jabar BA (2009) Different parameters for alkaloid detection in *Catharanthus roseus* tissue culture. Ibn Al-Haitham J Pure Appl Sci 22(2):44–50

Alam P, Khan ZA, Abdin MZ, Khan JA, Ahmad P, Elkholy SF, Sharaf-Eldin MA (2017) Efficient regeneration and improved sonication-assisted *Agrobacterium transformation* (SAAT) method for *Catharanthus roseus*. 3 Biotech 7:26

Al-Khateeb E, Alani NK, Jabar BA (2009) Different parameters for alkaloid detection in *Catharanthus roseus* tissue culture. Ibn Al-Haitham J Pure Appl Sci 22(2):1–7

Almagro L, López Perez AJ, Pedreño MA (2010) New method to enhance ajmalicine production in *Catharanthus roseus* cell cultures based on the use of cyclodextrins. Biotechnol Lett 33(2):381–385

Al-Oubaidi HKM, Amin ASM (2014) Effect of benzyl adenine on multiplication of Catharanthus roseus L. *in vitro*. World J Pharm Pharm Sci 3(6):2101–2107

Al-Zuhairi EMA, Ghanm NS (2017) Effect of jasmonic acid (JA) and glutamine on callus induction of Madagascar periwinkle plant (*Catharanthus roseus* L. cv. Nirvana Pink Blush) by *in vitro* culture. Int J Curr Microbiol App Sci 6(5):1415–1422

Angelova Z, Georgiev S, Roos W (2006) Elicitation of plants. Biotechnol Biotechnol Equip 20:72–83

Asada M, Shuler ML (1989) Stimulation of ajmalicine production and excretion from *Catharanthus roseus*: effects of adsorption in situ, elicitors and alginate immobilization. Appl Microbiol Biotechnol 30:475–481

Aslam J, Khan SH, Siddiqui ZH, Fatima Z, Maqsood M, Bhat MA, Nasim SA, Ilah A, Ahmad IZ, Khan SA, Mujib A, Sharma MP (2010) *Catharanthus roseus* (L.) G. Don. An important drug: its application and production. Pharm Glob IJCP 4:12

Aslam J, Mujib A, Nasim SA, Sharma MP (2009) Screening of vincristine yield in *ex vitro* and *in vitro* somatic embryos derived plantlets of *Catharanthus roseus* L. (G) Don. Sci Hortic 119:325–329

Aslam J, Mujib A, Sharma MP (2014) Somatic embryos in *Catharanthus roseus*: a scanning electron microscopic study. Not Sci Biol 6(2):167–172

Ataei-Azimi A, Hashemloian BD, Ebrahimzadeh H, Majd A (2008) High *in vitro* production of ant-canceric indole alkaloids from periwinkle (*Catharanthus roseus*) tissue culture. Afr J Biotechnol 7(16):2834–2839

Bachiri Y, Gazeau C, Hansz J, Morisset C, Dereuddre J (1995) Successful cryopreservation of suspension cells by encapsulation dehydration. Plant Cell Tissue Organ Cult 43:241–248

Baenas N, Garcia-Viguera C, Moreno DA (2014) Elicitation: a tool for enriching the bioactive composition of foods. Molecules 19(9):13541–13563

Bailey CM, Nicholson H (1990) Optimal temperature control for structured model of plant cell culture. Biotechnol Bioeng 35:252–258

Bakrudeen AAA, Shanthi GS, Gouthaman T, Kavitha MS, Rao MV (2011) *In vitro* micropropagation of *Catharanthus roseus* – an anticancer medicinal plant. Acta Bot Hungar 53(1–2):197–209

Batra J, Dutta A, Singh D, Kumar S, Sen J (2004) Growth and terpenoid indole alkaloid production in *Catharanthus roseus* hairy root clones in relation to left-and right-termini-linked Ri T-DNA gene integration. Plant Cell Rep 23:148–154

Begum T, Mathur M (2014) *In vitro* regeneration of *Catharanthus roseus* and *Bacopa monnieri* and their survey around Jaipur District. Int J Pure App Biosci 2(4):210–221

Benyammi R, Paris C, Khelifi-Slaoui M, Zaoui D, Belabbassi O, Bakiri N, Aci MM, Harfi B, Malik S, Makhzoum A, Desobry S, Khelifi L (2016) Screening and kinetic studies of catharanthine and ajmalicine accumulation and their correlation with growth biomass in *Catharanthus roseus* hairy roots. Pharm Biol 54(10):2033–2043. https://doi.org/10.3109/13880209.2016.1140213

Binder BYK, Peebles CAM, Shanks JV, San KY (2009) The effects of UV-B stress on the production of terpenoid indole alkaloids in *Catharanthus roseus* hairy roots. Biotechnol Prog 25(3):861–865

Blom TJM, Sierra M, Van Vliet TB, Franke-Van Dijk MEI, De Koning P, Van Iren F, Verpoorte R, Libbenga KR (1991) Uptake and accumulation of ajmalicine into isolated vacuoles of cultured cells of *Catharanthus roseus* (L.) G. Don and its conversion into serpentine. Planta 183:170–177

Boller T (1995) Chemoperception of microbial signals in plant cells. Annu Rev Plant Physiol Plant Mol Biol 46:189–214

Brodelius P, Nilsson K (1983) Permeabilization of immobilized plant cells, resulting in release of intracellularly stored products with preserved cell viability. Eur J Appl Microbiol Biotechnol 17:275–280

Campos-Tamayo F, Hernandez-domiguez E, Vazquez-flota F (2008) Vindoline formation in shoot cultures of *Catharanthus roseus* is synchronously activated with morphogenesis through the last biosynthetic step. Ann Bot 102:409–415

Canel C, Lopes-Cardoso MI, Whitmer S, Van der Fits L, Pasquali G, Van der Heijden R, Hoge JHC, Verpoorte R (1998) Effects of over-expression of strictosidine synthase and tryptophan decarboxylase on alkaloid production by cell cultures of *Catharanthus roseus*. Planta 205:414–419

Chen THH, Kartha KK, Constabel F, Gusta LV (1984a) Freezing characteristics of cultured *Catharanthus roseus* (L). G. Don cells treated with dimethylsulfoxide and sorbitol in relation to cryopreservation. Plant Physiol 75:720–725

Chen THH, Kartha KK, Leung NL, Kurz WGW, Chatson KB, Constabel F (1984b) Cryopreservation of alkaloid-producing cell cultures of periwinkle (Catharanthus roseus). Plant Physiol 75:726–731

Choi PS, Kim YD, Choi KM, Chung HJ, Choi DW, Liu JR (2004) Plant regeneration from hairy-root cultures transformed by infection with *Agrobacterium rhizogenes* in *Catharanthus roseus*. Plant Cell Rep 22:828–831

Contin A, Van der Heijden R, Hens JG, Hoopen HJG, Verpoorte R (1998a) The inoculum size triggers tryptamine or secologanin biosynthesis in a *Catharanthus roseus* cell culture. Plant Sci 139:205–211

Contin A, Van der Heijden R, Lefeber AW, Verpoorte R (1998b) The iridoid glucoside secologanin is derived from the novel triose phosphate/pyruvate pathway in a *Catharanthus roseus* cell culture. FEBS Lett 434:413–416

Contin A, Van der Heijden R, Verpoorte R (1999) Effects of alkaloid precursor feeding and elicitation on the accumulation of secologanin in a *Catharanthus roseus* cell suspension culture. Plant Cell Tissue Org Cult 56:111–119

Costa MMR, Hilliou F, Duarte P, Pereira LG, Almeida I, Leech M, Memelink J, Barcelo AR, Sottomayor M (2008) Molecular cloning and characterization of a vacuolar class III peroxidase involved in the metabolism of anticancer alkaloids in *Catharanthus roseus*. Plant Physiol 146:403–417

De Luca V, Marineau C, Brisson N (1989) Molecular cloning and analysis of cDNA encoding a plant tryptophan decarboxylase: comparison with animal dopa decarboxylases. PNAS 86(8):2582–2586

Decendit A, Liu D, Ouelhazi L, Doireau P, Merillon J, Rideau M (1992) Cytokinin-enhanced accumulation of indole alkaloids in *Catharanthus roseus* cell cultures-the factors affecting the cytokinin response. Plant Cell Rep 11:400–403

Dhandapani M, Kim DH, Hong SB (2008) Efficient plant regeneration via somatic embryogenesis and organogenesis from the explants of *Catharanthus roseus*. In Vitro Cell Dev Biol Plant 44:18–25

Eilert U, De Luca V, Constabel F, Kurz WGW (1987) Elicitor-mediated induction of tryptophan decarboxylase and strictosidine synthase activities in cell suspension cultures of *Catharanthus roseus*. Arch Biochem Biophys 254(2):491–497

El-Sayed M, Verpoorte R (2002) Effect of phytohormones on growth and alkaloid accumulation by a *Catharanthus roseus* cell suspension cultures fed with alkaloid precursors tryptamine and loganin. Plant Cell Tissue Organ Cult 68:265–270

El-Sayed M, Verpoorte R (2007) Catharanthus terpenoid indole alkaloids: biosynthesis and regulation. Phytochem Rev 6:277–305

El-Sayed M, Choi YH, Frederich M, Roytrakul S, Verpoorte R (2004) Alkaloid accumulation in *Catharanthus roseus* cell suspension cultures fed with stemmadenine. Biotechnol Lett 26:793–798

Faheem M, Singh S, Tanwer BS, Khan M, Shahzad A (2011) *In vitro* regeneration of multiplication shoots in *Catharanthus roseus*-an important medicinal plant. Adv Appl Sci Res 2(1):208–213

Farmer EE, Ryan CA (1990) Interplant communication: airborne methyl jasmonate induces synthesis of proteinase inhibitors in plant leaves. Proc Natl Acad Sci 87:7713–7716

Fathalla MA, Abd-El Kawy AM, Taha HS (2011) Effect of heavy metal ($HgCl_2$) on accumulation and production of total indole alkaloids, vinblastine and/or vincristine from Egyptian *Catharanthus Roseus* (L.) G. Don. calli cultures. J Appl Sci Res 7(4):542–549

Fatima S, Mujib A, Nasim SA, Siddiqui ZH (2009) Cryopreservation of embryogenic cell suspensions of *Catharanthus roseus* L. (G) Don. Plant Cell Tissue Organ Cult 98:1–9

Fatima S, Mujib A, Dipti T (2015) NaCl amendment improves vinblastine and vincristine synthesis in *Catharanthus roseus*: a case of stress signalling as evidenced by antioxidant enzymes activities. Plant Cell Tissue Organ Cult 121(2):445–458

Fulzele DP, Heble MR (1994) Large-scale cultivation of *Catharanthus roseus* cells: production of ajmalicine in a 20-1 airlift bioreactor. J Biotechnol 35:1–7

Gamborg OL, Shyluk JP (1981) Nutrition, media and characteristics of plant cell and tissue cultures. In: Thorpe TA (ed) Plant tissue culture: methods and applications in agriculture. Academic Press, New York, pp 21–44

Garnier F, Label P, Hallard D, Chenieux JC, Rideau M, Hamdi S (1996) Transgenic periwinkle tissues overproducing cytokinins do not accumulate enhanced levels of indole alkaloids. Plant Cell Tissue Organ Cult 45(3):223–230

Gaviraj EN, Veeresham C (2006) Effect of precursors and organic compounds on alkaloid production in transformed root cultures of *Catharanthus roseus* var. nirmal. Pharm Biol 44(5):371–377

Giri A, Narasu ML (2000) Transgenic hairy roots. Recent trends and applications. Biotechnol Adv 18:1–22

Godoy-Hernandez GC, Loyola-Vargas VM (1991) Effect of fungal homogenates, enzyme inhibitors and osmotic stress on alkaloids content of *Catharanthus roseus* cell suspension cultures. Plant Cell Rep 10:537–540

Godoy-Hernandez GC, Vazquez-Flota FA, Loyola-Vargas VM (2000) The exposure to trans-cinnamic acid of osmotically stressed *Catharanthus roseus* cells cultured in a 14-L bioreactor increases alkaloid accumulation. Biotechnol Lett 22(11):921–925

Goklany S, Loring RH, Glick J, Lee-Parsons CWT (2009) Assessing the limitations to terpenoid indole alkaloid biosynthesis in *Catharanthus roseus* hairy root cultures through gene expression profiling and precursor feeding. Biotechnol Prog 25(5):1289–1296

Goldhaber-Pasillas GD, Mustafa NR, Verpoorte R (2014) Jasmonic acid effect on the fatty acid and terpenoid indole alkaloid accumulation in cell suspension cultures of *Catharanthus roseus*. Molecules 19:10242–10260

Gorelick J, Bernstein N (2014) Chapter five—Elicitation: an underutilized tool in the development of medicinal plants as a source of therapeutic secondary metabolites. In: Sparks DL (ed) Advances in agronomy, vol 124. Elsevier, Amsterdam, The Netherlands, pp 201–230

Guirimand G, Courdavault V, Lanoue A, Mahroug S, Guihur A, Blanc N, Giglioli-Guivarc'h N, St-Pierre B, Burlat V (2010) Strictosidine activation in apocynaceae: towards a "nuclear time bomb"? BMC Plant Biol 10:182–201

Guirimand G, Guihur A, Poutrain P, Héricourt F, Mahroug S, St-Pierre B, Burlat V, Courdavault V (2011) Spatial organization of the vindoline biosynthetic pathway in *Catharanthus roseus*. J Plant Physiol 168:549–557

Hanafy MS, Matter MA, Asker MS, Rady MR (2016) Production of indole alkaloids in hairy root cultures of *Catharanthus roseus* L. and their antimicrobial activity. S Afr J Bot 105:9–18

He L, Yang LI, Tan R, Zhao S, Hu Z (2011) Enhancement of vindoline production in suspension culture of the *Catharanthus roseus* cell line C20hi by light and methyl jasmonate elicitation. Anal Sci 27:1243–1248

He S, Zhu J, Zi J, Zhou P, Liang J, Yu R (2015) A novel terpenoid indole alkaloid derived from catharanthine via biotransformation by suspension-cultured cells of *Catharanthus roseus*. Biotechnol Lett 37:2481–2487

Hernandez-Dominguez E, Campos-Tamayo F, Vazquez-Flota F (2004) Vindoline synthesis in *in vitro* shoot cultures of *Catharanthus roseus*. Biotechnol Lett 26:671–674

Hirata K, Yamanaka A, Kurano N, Miyamoto K, Miura Y (1987) Production of indole alkaloids in multiple shoot culture of *Catharanthus roseus* (L). G. Don. Agric Biol Chem 51(5):1311–1317

Ho CH, Shanks JV (1992) Effects of initial medium pH on growth and metabolism of *Catharanthus roseus* hairy root cultures: a study with ^{31}P and ^{13}C NMR spectroscopy. Biotechnol Lett 14(10):959–964

Hong SB, Hughes EH, Shanks JV, San KY, Gibson SI (2003) Role of the non-mevalonate pathway in indole alkaloid production by *Catharanthus roseus* hairy roots. Biotechnol Prog 19:1105–1108

Hoopen HJG, Gulik WM, Schlatmann JE, Moreno PRH, Vinke JL, Heijnen JJ, Verpoorte R (1994) Ajmalicine production by cell cultures of *Catharanthus roseus*: from shake flask to bioreactor. Plant Cell Tissue Organ Cult 38:85–91

Hughes EH, Shanks JV (2002) Metabolic engineering of plants for alkaloid production. Metab Eng 4(1):41–48

Iwase A, Aoyagi H, Ohme-takagi M, Tanaka H (2005) Development of a novel system for producing ajmalicine and serpentine using direct culture of leaves in *catharanthus roseus* intact plant. J Biosci Bioeng 99(3):208–215

Jeong CS, Chakrabarty D, Hahn EJ, Lee HL, Paek KY (2006) Effects of oxygen, carbon dioxide and ethylene on growth and bioactive compound production in bioreactor culture of ginseng adventitious roots. Biochem Eng J 27:252–263

Junaid A, Mujib A, Bhat MA, Sharma MP (2006) Somatic embryo proliferation, maturation and germination in *Catharanthus roseus*. Plant Cell, Tissue Organ Cult 84:325–332

Junaid A, Mujib A, Bhat MA, Sharma MP, Šamaj J (2007) Somatic embryogenesis and plant regeneration in *Catharanthus roseus*. Biol Plant 51(4):641–646

Junaid A, Mujib A, Sharma MP (2011) Influence of freezing and non-freezing temperature on somatic embryogenesis and vinblastine production in *Catharanthus roseus* (L.) G. Don. Acta Physiol Plant 33:473–480

Jung KH, Kwak SS, Kim SW, Lee H, Choi CY, Liu JR (1992) Improvement of the catharanthine productivity in hairy root cultures of *Catharanthus roseus* by using monosaccharides as a carbon source. Biotechnol Lett 14(8):695–700

Kalidass C, Mohan VR, Daniel A (2010) Effect of auxin and cytokinin on vincristine production by callus cultures of *Catharanthus roseus* L. (Apocynaceae). Trop Subtrop Agroecosyst 12:283–288

Kartha KK (1981) Meristem culture and cryopreservation methods and applications. In: Thorpe TA (ed) Plant tissue culture: methods and applications in agriculture. Academic Press, New York, pp 181–211

Kartha KK, Leung NL, Gaudet-LaPrarire P, Constable F (1982) Cryopreservation of periwinkle, *Catharanthus roseus* cells cultured in vitro. Plant Cell Rep 1(4):135–138

Kashyab S, Kale RD (2015) *In vitro* tissue culture of *Catharanthus roseus* using vermicompost extract and coelomic fluid-an innovative and novel approach. Int J Curr Res 7(12):24679–24683

Keasling JD (2008) Synthetic biology for synthetic chemistry. ACS Chem Biol 3:64–76

Khashan KT, Al-Athary MAH (2016) Vinblastine and vincristine alkaloids production from callus of *Catharanthus roseus* (L.) G. Don under some abiotic factors. Al-Kufa Univ J Biol 8(2):75–90

Khashan KT, Husain MA (2015) Effect of biotic factors stresses on vinblastine and vincristine production from callus of *Catharanthus roseus*. Euphrates J Agric Sci 7(2):25–41

Knobloch KH, Berlin J (1980) Influence of medium composition on the formation of secondary compounds in cell suspension cultures of *Catharanthus roseus* (L.) G. Don. Z Naturforsch 35:551–556

Knobloch KH, Berlin J (1983) Influence of phosphate on the formation of the indole alkaloids and phenolic compounds in cell suspension cultures of *Catharanthus roseus*. I. Comparison of enzyme activities and product accumulation. Plant Cell Tissue Organ Cult 2:333–340

Kumar SR, Shilpashree HB, Nagegowda DA (2018) Terpene moiety enhancement by overexpression of geranyl (geranyl) diphosphate synthase and geraniol synthase elevates monomeric and dimeric monoterpene indole alkaloids in transgenic *Catharanthus roseus*. Front Plant Sci 9:942

Kurz WGW, Chatson KB, Constabel F, Kutney JP, Choi LSL, Kolodziejczyk P, Sleigh SK, Stuart KL, Worth BR (1981) Alkaloid production in *Catharanthus roseus* cell cultures VIII. Planta Med 42(5):22–31

Lee-Parsons CWT, Ertürk S, Tengtrakool J (2004) Enhancement of ajmalicine production in *Catharanthus roseus* cell cultures with methyl jasmonate is dependent on timing and dosage of elicitation. Biotechnol Lett 26:1595–1599

Li M, Peebles CAM, Shanks JV, San KY (2011) Effect of sodium nitroprusside on growth and terpenoid indole alkaloid production in *Catharanthus roseus* hairy root cultures. Biotechnol Prog 27(3):625–630

Liang C, Chen C, Zhou P, Xu L, Zhu J, Liang J, Zi J, Yu R (2018) Effect of *Aspergillus flavus* fungal elicitor on the production of terpenoid indole alkaloids in *Catharanthus roseus* cambial meristematic cells. Molecules 23(3276):1–16

Limam F, Chahed K, Ouelhazi N, Ghrir R, Ouelhazi L (1998) Phytohormone regulation of isoperoxidases in *Catharanthus roseus* suspension cultures. Phytochemistry 49(5):1219–1225

Lloyd G, McCown BH (1980) Commercially-feasible micropropagation of mountain laurel, Kalmia latifolia, by use of shoot-tip culture. Combi Proc Int Plant Propag Soc 30:421–427

Makhzoum A, Bjelica A, Petit-Paly G, Mark A, Bernards MA (2015) Novel plant regeneration and transient gene expression in *Catharanthus roseus*. All Res J Biol 6:1–9

Malabadi RB, Mulgund GS, Nataraja K (2009) Triacontanol induced somatic embryogenesis and plantlet regeneration in *Catharanthus roseus*. J Med Aromat Plant Sci 31:147–151

Mandagi SJM, Pandiangan D, Tilaar W (2017) Content of ajmalicine on cultured callus *Catharanthus roseus* (L.) G. Don with tryptophan treatment. Int J Sci Res 6(5):1919–1922

Mannonen L, Toivonen L, Kauppinen VC (1990) Effects of long term preservation on growth and productivity of *Panax ginseng* and *Catharanthus roseus* cell culture. Plant Cell Rep 9:173–177

Maqsood M, Abdul M (2017) Yeast extract elicitation increases vinblastine and vincristine yield in protoplast derived tissues and plantlets in *Catharanthus roseus*. Braz J Pharmacogn 27:549–556

Marsh Z, Yang T, Nopo-Olazabal L, Wu S, Ingle T, Joshee N, Medina-Bolivar F (2014) Effect of light, methyl jasmonate and cyclodextrin on production of phenolic compounds in hairy root cultures of *Scutellaria lateriflora*. Phytochemistry 107:50–60

Meijer AH, Lopes Cardoso ML, Voskuilen JT, De Waal A, Verpoorte R, Hoge JH (1993) Isolation and characterization of a cDNA clone from *Catharanthus roseus* encoding NADPH: cytochrome P-450 reductase, an enzyme essential for reactions catalysed by cytochrome P-450 mono-oxygenases in plants. Plant J 4:47–56

Mekky H, Al-Sabahi J, Abdel-Kreem MFM (2018) Potentiating biosynthesis of the anticancer alkaloids vincristine and vinblastine in callus cultures of *Catharanthus roseus*. S Afr J Bot 114:29–31

Menke FLH, Parchmann S, Mueller MJ, Kijne JW, Memelink J (1999) Involvement of the octadecanoid pathway and protein phosphorylation in fungal elicitor induced expression of terpenoid indole alkaloid biosynthetic genes in *Catharanthus roseus*. Plant Physiol 119:1289–1296

Mérillon JM, Doireau P, Guillot A, Chénieux JC, Rideau M (1986) Indole alkaloid accumulation and tryptophan decarboxylase activity in *Catharanthus roseus* cells cultured in three different media. Plant Cell Rep 1:23–26

Mishra MR, Srivastava RK, Akhtar N (2018a) Enhanced alkaloid production from cell culture system of *Catharanthus roseus* in combined effect of nutrient salts, sucrose and plant growth regulators. J Biotechnol Biomed Sci 1(4):14–34

Mishra MR, Srivastava RK, Akhtar N (2018b) Enhancing alkaloid production from cell culture system of *Catharanthus roseus* with different carbon sources. Eur J Biotechnol Biosci 6(5):12–20

Mitra A, Khan BM, Rawal SK (1997) Photoautotrophic shoot culture-an economical alternative for the production of total alkaloid from *Catharanthus roseus* (l.) G. Don. Curr Sci 73(7):608–609

Miura Y, Hirata K (1987) Isolation of vinblastine in callus culture with differentiated roots of *Catharanthus roseus* (L). G. Don. Agric Biol Chem 51(2):611–614

Moghe S, Laud D, Dehankar B, Moghe R, Sadhu P, Ade G, Ishani B, Hadke A (2016) Effect of growth regulator combination on *in-vitro* regeneration of *Catharanthus roseus*. Int J Life Sci Special Issue A6:1–4

Moreno PRH, Schlatmann JE, Van der Heijden R, Van Gulik WM, Hoopen HJT, Verpoorte R, Heijnen JJ (1993a) Induction of ajmalicine formation and related enzyme activities in *Catharanthus roseus* cells: effect of inoculum density. Appl Microbiol Biotechnol 39:42–47

Moreno PRH, Van der Heijden R, Verpoote R (1993b) Effect of terpenoid precursor feeding and elicitation on formation of indole alkaloid in cell suspension cultures in *Catharanthus roseus*. Plant Cell Rep 12:702–705

Moreno PRH, Van der Heijden R, Verpoorte R (1995) Cell and tissue cultures of *Catharanthus roseus*; a literature survey II. Updating from 1988-1993. Plant Cell Tissue Organ Cult 42:1–25

Morgan JA, Shanks JV (1999) Inhibitor studies of tabersonine metabolism in *C. roseus* hairy roots. Phytochemistry 51:61–68

Morgan JA, Shanks JV (2000) Determination of metabolic rate-limitations by precursor feeding in *Catharanthus roseus* hairy root cultures. J Biotechnol 79:137–145

Morgan JA, Barney CS, Penn AH, Shanks V (2000) Effects of buffered media upon growth and alkaloid production of *Catharanthus roseus* hairy roots. Appl Microbiol Biotechnol 53:262–265

Morris P (1986) Regulation of product synthesis in cell cultures of *Catharanthus roseus*. Effect of culture temperature. Plant Cell Rep 5:427–429

Mujib A, Ilah A, Aslam J, Fatima S, Siddiqui ZH, Maqsood M (2012) *Catharanthus roseus* alkaloids: application of biotechnology for improving yield. Plant Growth Regul 68:111–127

Mujib A, Ali M, Isah T, Tonk D (2014) Somatic embryo mediated mass production of *Catharanthus roseus* in culture vessel (bioreactor)–a comparative study. Saudi J Biol Sci 21:442–449

Murashige T, Skoog F (1962) A revised medium for rapid growth and bioassays with tobacco tissue cultures. Physiol Plant 15:473–497

Muthukumar B, Arochiasamy DI, Natarajan E (2004) Direct organogenesis in *Datura metel* L. from *in vitro* and *in vivo* nodal explants. Indian J Biotechnol 3:449–451

Negi RS (2011) Fast *in-vitro* callus induction in *Catharanthus roseus* – a medicinally important plant used in cancer therapy. Res J Pharm Biol Chem Sci 2(4):597–603

Nitsch JP, Nitsch C (1969) Haploid plants from pollen grains. Science 163:85–87

Noormohammadi Z, Taban M, Farahani F (2018) The impact of gamma radiation on *Tdc* and *Str* gene expressions in *Catharanthus roseus* regenerated plantlets. Biodiversitas 19(5):1805–1810

Pan Q, Mustafa NR, Tang K, Choi YH, Verpoorte R (2016) Monoterpenoid indole alkaloids biosynthesis and its regulation in *Catharanthus roseus:* a literature review from genes to metabolites. Phytochem Rev 15:221–250

Pandiangan D, Nainggolan N (2006) The enhancement of catharanthine content in *Catharanthus roseus* callus culture treated with naphtalene acetic acid. Hayati 13(3):90–94

Pandiangan D, Tilaara W, Karyonob Esyantic RR, Subarnasb A (2010) The effect tryptophan feeding on growth, protein content and TDC activity of Catharanthus roseus (L) G. DON cell aggregate culture in the airlift bioreactor. In: Proceedings of the third international conference on mathematics and natural sciences (ICMNS 2010)

Pandiangan D, Tilaar W, Nainggolan N (2013) Morphological changes of cell in relation to increased catharanthine content of *Catharanthus roseus* cell aggregate culture after tryptophan treatment. Int J Basic Appl Sci 13(1):45–51

Pandiangan D, Tilaar W, Nainggolan N, Wahyudi L (2015) Relations between catharanthine content enhancement with the other associated secondary metabolites in *Catharanthus roseus* cell culture that treated tryptophan. Int J Sci Res 4(11):2208–2212

Papon N, Bremer J, Vansiri A, Andreu F, Rideau M, Creche J (2005) Cytokinin and ethylene control indole alkaloid production at the level of the MEP/terpenoid pathway in *Catharanthus roseus* suspension cells. Planta Med 71:572–574

Parr AJ, Peerless ACJ, Hamill JD, Walton NJ, Robins RJ, Rhodes MJC (1988) Alkaloid production by transformed root cultures of *Catharanthus roseus*. Plant Cell Rep 7:309–312

Pasquali G, Goddijin OJM, De Wall A, Verpoorte R, Schilperoort RA, Hoge JHC (1992) Coordinated regulation of two indole alkaloid biosynthetic genes from *Catharanthus roseus* by auxin and elicitors. Plant Mol Biol 18:1121–1131

Peebles CAM, Hong S, Gibson SI, Shanks JV (2006) Effects of terpenoid precursor feeding on *Catharanthus roseus* hairy roots over-expressing the alpha or the alpha and beta subunits of anthranilate synthase. Biotechnol Bioeng 93(3):534–540

Peebles CAM, Hughes EH, Shanks JV, San KY (2009) Transcriptional response of the terpenoid indole alkaloid pathway to the overexpression of ORCA3 a long with jasmonic acid elicitation of *Catharanthus roseus* hairy roots over time. Metab Eng 11:76–86

Perassolo M, Smith ME, Giulietti AM, Talou JR (2016) Synergistic effect of methyl jasmonate and cyclodextrins on anthraquinone accumulation in cell suspension cultures of *Morinda citrifolia* and *Rubia tinctorum*. Plant Cell Tissue Organ Cult 124(2):319–330

Pietrosiuk A, Furmanowa M, Łata B (2007) *Catharanthus roseus*: micropropagation and *in vitro* techniques. Phytochem Rev 6:459–473

Pliankong P, Suksa-Ard P, Wannakrairoj S (2018) Chitosan elicitation for enhancing of vincristine and vinblastine accumulation in cell culture of *Catharanthus roseus* (L.) G. Don. J Agric Sci 10(12):287–293

Pomahacova B, Dusek J, Duskova J, Yazaki K, Roytrakul S, Verpoorte R (2009) Improved accumulation of ajmalicine and tetrahydroalstonine in Catharanthus cells expressing an ABC transporter. J Plant Physiol 166:1405–1412

Rahmatzadeh S, Khara J, Kazemitabar SK (2014) The study of *in vitro* regeneration and growth parameters in *Catharanthus roseus* L. under application of tryptophan. J Sci Kharazmi Univ 14(3):249–260

Ramani S, Chelliah J (2007) UV-B-induced signaling events leading to enhanced-production of catharanthine in *Catharanthus roseus* cell suspension cultures. BMC Plant Biol 7:61

Raouf Fard F, Moieni A, Omidbaigi R (2008) Effects of different concentrations of α-naphthaleneacetic acid and 6-benzylaminopurine on shoot regeneration of *Vinca minor* L. J Agric Sci Technol 10:337–344

Rashmi R, Trivedi MP (2015) Rapid in-vitro regeneration of an important medicinal and an ornamental plant (Catharanthus roseus L). Biochem Anal Biochem 4(4):1–6

Ren AZ, Gao YB, Liu S (2002) Response of some protective enzymes in *Brassica chinensis* seedling to Pb^{++}, Cd^{+} and Cr^{+++} stress. Chin J Appl Ecol 13:510–512

Rijhwani SK, Shanks JV (1998) Effect of subculture cycle on growth and indole alkaloid production by *Catharanthus roseus* hairy root cultures. Enzym Microb Technol 22:606–611

Rizvi NF, Weaver JD, Cram EJ, Lee-Parsons CWT (2016) Silencing the transcriptional repressor, zct1, illustrates the tight regulation of terpenoid indole alkaloid biosynthesis in *Catharanthus roseus* hairy roots. PLoS One 11(7):1–17

Rodrigueza S, Compagnonb V, Crouchc NP, St-Pierred B, De Lucae V (2003) Jasmonate-induced epoxidation of tabersonine by a cytochrome P-450 in hairy root cultures of *Catharanthus roseus*. Phytochemistry 64:401–409

Ruiz-May E, Galaz-Avalos RM, Loyola-Vargas VM (2009) Differential secretion and accumulation of terpene indole alkaloids in hairy roots of *Catharanthus roseus* treated with methyl jasmonate. Mol Biotechnol 41:278–285

Saifullah, Khan S (2011) Callus induction and cell suspension culture production of *Catharanthus roseus* for biotransformation studies of (−)- caryophyllene oxide. Pak J Bot 43(1):467–473

Saiman MZ, Miettinen K, Mustafa NR, Choi YH, Verpoorte R, Schulte AE (2018) Metabolic alteration of *Catharanthus roseus* cell suspension cultures overexpressing *geraniol synthase* in the plastids or cytosol. Plant Cell Tissue Organ Cult 134:41–53

Schiel O, Berlin J (1987) Large scale fermentation and alkaloid production of cell suspension cultures of *Catharanthus roseus*. Plant Cell Tissue Organ Cult 8:153–161

Senoussi MM, Crèche J, Rideau M (2007) Relation between hypoxia and alkaloid accumulation in *Catharanthus roseus* cell suspension. J Appl Sci Res 3(4):287–290

Sharma A, Verma P, Mathur A, Mathur AK (2018) Genetic engineering approach using early *Vinca* alkaloid biosynthesis genes led to increased tryptamine and terpenoid indole alkaloids biosynthesis in differentiating cultures of *Catharanthus roseus*. Protoplasma 255:425–435

Sharma A, Mathur AK, Ganpathy J, Joshi B, Patel P (2019) Effect of abiotic elicitation and pathway precursors feeding over terpenoid indole alkaloids production in multiple shoot and callus cultures of *Catharanthus roseus*. Biologia 74:1–11

Sim SJ, Chang HN, Liu JR, Jung KH (1994) Production and secretion of indole alkaloids in hairy root cultures of *Catharanthus roseus:* effects of *in situ* adsorption, fungal elicitation and permeabilization. J Ferment Bioeng 78(3):229–234

Singh M, Sharma R, Banerjee UC (2002) Biotechnological applications of cyclodextrins. Biotechnol Adv 20(5–6):341–359

Singh R, Kharb P, Rani K (2011) Rapid micropropagation and callus induction of *Catharanthus roseus in vitro* using different explants. World J Agric Sci 7(6):699–704

Sivakumar G (2006) Bioreactor technology: a novel industrial tool for high-tech production of bioactive molecules and biopharmaceuticals from plant roots. Biotechnol J 1:1419–1427

Smeets K, Ruytinx J, Semane B, Belleghem F, Remans T, Sanden S, Vangronsvelda J, Cuypersa A (2008) Cadmium-induced transcriptional and enzymatic alterations related to oxidative stress. Environ Exp Bot 63:1–8

Srivastava S, Srivastava AK (2012) Strategies to overcome oxygen transfer limitations during hairy root cultivation of *Azadirachta indica* for enhanced azadirachtin production. Appl Biochem Biotechnol 167:1818–1830

Sun J, Manmathan H, Sun C, Peebles CAM (2016) Examining the transcriptional response of overexpressing anthranilate synthase in the hairy roots of an important medicinal plant *Catharanthus roseus* by RNA-seq. BMC Plant Biol 16:108

Swanberg A, Dai W (2008) Plant regeneration of periwinkle (Catharanthus roseus) via organogenesis. HortScience 43(3):832–836

Taha HS, El-Bahr MK, Seif-El-Nasr MM (2008) *In Vitro* studies on Egyptian *Catharanthus Roseus* (L.) G. Don.: 1-Calli production, direct shootlets regeneration and alkaloids determination. J Appl Sci Res 4(8):1017–1022

Taha HS, El-Bahr MK, Seif-El-Nasr MM (2009) *In vitro* studies on Egyptian *Catharanthus roseus* (L.) G.Don. IV: manipulation of some amino acids as precursors for enhanced of indole alkaloids production in suspension cultures. Aust J Basic Appl Sci 3(4):3137–3144

Taha HS, Shams KA, Nazif NM, Seif-El-Nasr MM (2014) *In vitro* studies on Egyptian *Catharanthus roseus* (L.) G.Don V: impact of stirred reactor physical factors on achievement of cells proliferation and vincristine and vinblastine accumulation. Res J Pharm Biol Chem Sci 5(2):330–340

Tanabe S, Hayashi N, Nishizawa Y, Yamane H, Shibuya N, Minami E (2008) Elicitor and calatalse activity of conidia suspensions of various strains of *Magnaporthe grisea* in suspension-cultured cells of rice. Biosci Biotechnol Biochem 72:889–892

Thakore D, Srivastava AK, Sinha AK (2013) Yield enhancement strategies for enhancement of indole alkaloids in hairy root cultures of *Catharanthus roseus*. Int J Chem Eng Appl 4(3):153–156

Thakore D, Srivastava AK, Sinha AK (2015) Model based fed batch cultivation and elicitation for the overproduction of ajmalicine from hairy roots of *Catharanthus roseus*. Biochem Eng J 97:73–80

Thakore D, Srivastavaa AK, Sinhab AK (2017) Mass production of ajmalicine by bioreactor cultivation of hairy roots of *Catharanthus roseus*. Biochem Eng J 119:84–91

Tikhomiroff C, Allais S, Klvana M, Hisiger S, Mario Jolicoeur M (2002) Continuous selective extraction of secondary metabolites from *Catharanthus roseus* hairy roots with silicon oil in a two-liquid-phase bioreactor. Biotechnol Prog 18:1003–1009

Toivonen L, Ojala M, Kauppinenb V (1991) Studies on the optimization of growth and lndole alkaloid production by hairy root cultures of *Catharanthus roseus*. Biotechnol Bioeng 37:673–680

Toivonen L, Laakso S, Rosenqvist H (1992) The effect of temperature on hairy root cultures of *Catharanthus roseus*: growth, indole alkaloid accumulation and membrane lipid composition. Plant Cell Rep 11:395–399

Tonk D, Mujib A, Maqsood M, Ali M, Zafar N (2016) *Aspergillus flavus* fungus elicitation improves vincristine and vinblastine yield by augmenting callus biomass growth in *Catharanthus roseus*. Plant Cell Tissue Organ Cult 126:291–303

Van Der Heijden R, Jacobs DI, Snoeijer W, Hallard D, Verpoorte R (2004) The catharanthus alkaloids: pharmacognosy and biotechnology. Curr Med Chem 11(5):607–628

Veerabathini S, Sarang S, Shalini S, Deepa Sankar P (2015) Standardization of friable callus development in *Catharanthus roseus* (Linn.) G. DON. Int J Pharm Pharm Sci 7(3):111–113

Verma P, Mathur AK (2011a) *Agrobacterium* tumefaciens-mediated transgenic plant production via direct shoot bud organogenesis from pre-plasmolyzed leaf explants of *Catharanthus roseus*. Biotechnol Lett 33:1053–1060

Verma P, Mathur AK (2011b) Direct shoot bud organogenesis and plant regeneration from pre plasmolysed leaf explants in *Catharanthus roseus*. Plant Cell Tissue Organ Cult 106:401–408

Verma AK, Singh RR, Singh S (2012) Improved alkaloid content in callus culture of *Catharanthus roseus*. Bot Serb 36(2):123–130

Verma P, Khan SA, Mathur AK, Shanker K, Kalra A (2014) Fungal endophytes enhanced the growth and production kinetics of *Vinca minor* hairy roots and cell suspensions grown in bioreactor. Plant Cell Tissue Organ Cult 118:257–268

Verma P, Sharma A, Khan SA, Shanker K, Mathur AK (2015) Over-expression of *Catharanthus roseus* tryptophan decarboxylase and strictosidine synthase in *rol* gene integrated transgenic cell suspensions of *Vinca minor*. Protoplasma 252:373–381

Verpoorte R, Van der Heijden R, Van Gulik WM, Ten Hoopen HJG (1991) Plant biotechnology for the production of alkaloids present status and prospects. In: Brossi A (ed) The alkaloids, vol 40. Academic Press, San Diego, pp 1–187

Wang Q, Xing S, Pan Q, Yuan F, Zhao J, Tian Y, Chen Y, Wang G, Tang K (2012) Development of efficient *Catharanthus roseus* regeneration and transformation system using agrobacterium tumefaciens and hypocotyls as explants. BMC Biotechnol 12:34

Whitmer S, Canel C, Hallard D, Goncalves C, Verpoorte R (1998) Influence of precursor availability on alkaloid accumulation by transgenic cell line of *Catharanthus roseus*. Plant Physiol 116:853–857

Whitmer S, Van der Heijn R, Verpoorte R (2002) Effect of precursor feeding on alkaloid accumulation by a tryptophan decarboxylase over-expressing transgenic cell line T22 of *Catharanthus roseus*. J Biotechnol 96(2):193–203

Whitmer S, Canel C, Van der Heijden R, Verpoorte R (2003) Long-term instability of alkaloid production by stably transformed cell lines of *Catharanthus roseus*. Plant Cell Tissue Organ Cult 74(1):73–80

Xiang BB, Zhu YR, Wang WJ, Bai YL, Wang Y (2011) Cell line screening of *Catharanthus roseus* for high yield production of ajmalicine. J Med Plant Res 5(3):420–424

Yang L, Stöckigt J (2010) Trends for diverse production strategies of plant medicinal alkaloids. Nat Prod Rep 27(10):1469–1479

Yuan F, Wang Q, Pan Q, Wang G, Zhao J, Tian Y, Tang K (2011) An efficient somatic embryogenesis based plant regeneration from the hypocotyl of *Catharanthus roseus*. Afr J Biotechnol 10(66):14786–14795

Zargar M, Farahani F, Nabavi T (2010) Hairy roots production of transgenic *Catharanthus roseus* L. plants with *Agrobacterium rhizogenes* under *in vitro* conditions. J Med Plant Res 4(21):2199–2203

Zhao J, Verpoorte R (2007) Manipulating indole alkaloid production by *Catharanthus roseus* cell cultures in bioreactors: from biochemical processing to metabolic engineering. Phytochem Rev 6:435–457

Zhao J, Zhu WH, Hu Q (2000) Enhanced ajmalicine production in *Catharanthus roseus* cell cultures by combined elicitor treatment: from shake-flask to 20-l airlift bioreactor. Biotechnol Lett 22:509–514

Zhao J, Hu Q, Guo YQ, Zhu WH (2001a) Effects of stress factors, bioregulators, and synthetic precursors on indole alkaloid production in compact callus clusters cultures of *Catharanthus roseus*. Appl Microbiol Biotechnol 55:693–698

Zhao J, Zhu WH, Hu Q, Guo YQ (2001b) Compact callus cluster suspension cultures of *Catharanthus roseus* with enhanced indole alkaloid biosynthesis. In Vitro Cell Dev Biol Plant 37:68–72

Zhao J, Zheng B, Li Y, Shan T, Mou Y, Lu S, Li P, Zhou L (2011) Enhancement of diepoxin ζ production by yeast extract and its fractions in liquid culture of Berkleasmium-like endophytic fungus Dzf12 from *Dioscorea zingiberensis*. Molecules 16:847–856

Zhao L, Sander GW, Shanks JV (2013) Perspectives of the metabolic engineering of terpenoid indole alkaloids in *Catharanthus roseus* hairy roots. Adv Biochem Eng Biotechnol 134:23–54

Zhen-Gui Z, Di L, Zhi-Bi H (1998) Comparison of cell growth and alkaloid production of *Catharanthus roseus* cells cultured in shake flask and in bioreactor. Acta Bot Sin 40(1):51–55

Zhong JJ (2001) Biochemical engineering of the production of plant-specific secondary metabolites by cell suspension cultures. Adv Biochem Eng Biotechnol 72:1–26

Zhong JJ, Pan ZW, Wang ZY, Wu J, Chen YF, Takagi M, Yoshida T (2002) Effect of mixing time on taxoid production using suspension cultures of *Taxus chinensis* in a centrifugal impeller bioreactor. J Biosci Bioeng 94:244–250

Zhou P, Yang J, Zhu J, He S, Zhang W, Yu R, Zi J, Song L, Huang X (2015) Effects of β-cyclodextrin and methyl jasmonate on the production of vindoline, catharanthine, and ajmalicine in *Catharanthus roseus* cambial meristematic cell cultures. Appl Microbiol Biotechnol 99(17):7035–7045

Zhu X, Zeng X, Sun C, Shilin Chen S (2014) Biosynthetic pathway of terpenoid indole alkaloids in *Catharanthus roseus*. Front Med 8(3):285–293

Zhu J, Wang M, Wen W, Yu R (2015) Biosynthesis and regulation of terpenoid indole alkaloids in *Catharanthus roseus*. Pharmacogn Rev 9:24–28

Zulkepli AZ, Samad AA (2011) Optimization of sterilization method, callus and shoot induction of *Catharanthus roseus* explants. Int J Biosci 1(1):31–35

Chapter 2
Plant Biotechnology and Milk Thistle

Contents

M. R. Rady, *Plant Biotechnology and Medicinal Plants*,
https://doi.org/10.1007/978-3-030-22929-0_2

Abstract *Silybum marianum* is one of the most extensively studied medicinal herbs with well-known hepatoprotective activity. This chapter gives an overview of the various approaches used to optimize induction of cell, suspension, organ, root, and transgenic cultures from *S. marianum*. Additionally, examples of using these different approaches are shown and discussion for the production of silymarin from *S. marianum* tissue culture is also detailed. This chapter summarizes the recent research work of various in vitro culture, abiotic and biotic elicitors, and precursor feeding applied to *S. marianum* cultural system and their stimulating effects on the accumulation of silymarin.

Keywords *Silybum marianum* · Biosynthesis · Cell and callus culture · Regeneration · Root cultures · Transformation · Hairy root cultures · Elicitation · Flavonolignans

1 Introduction

Silybum marianum L. The *Silybum marianum* (L.) Gaertn. is the dicotyledonous herbs of the Asteraceae family that is important in medical industry. Milk thistle is native to the Mediterranean area but has been grown for centuries throughout Europe. It also grows in India, China, Africa, and Australia (Corchete 2008). Its extracts are being used as antihepatotoxic therapy for liver diseases. The biological active compound of *S. marianum* is a mixture of several flavonolignans generally known as silymarin (Eari et al. 2017). Silymarin is a strong antioxidant that has been proved to promote liver cell regeneration (Sagar 2007). Silymarin is a polyphenolic flavonoid mixture isolated from milk thistle which is believed to be responsible for the plant's hepatoprotective action (El Sherif et al. 2013). However, silymarin is composed of a mixture of the flavonolignans, silychristin (SC), isosilychristin (ISC), silydianin (SD), silybin (SB) A and B, and isosilybin (ISB) A and B with the precursor flavonoid taxifolin (TXF) (Kim et al. 2003; Lee and Liu 2003; Khalili et al. 2009). It was reported that average sale of silymarin is about US$ 8 billion/annum and its demand varies from 18 to 20 tons/year (Khan et al. 2013). Silybum plants have many agricultural problems which cause reduction of the total yield and quality of silymarin. Breeding studies for this plant are very low due to difficulties in its morphological features, like strong thorny stem, spiked leaves, and flowers tipped (Fig. 2.1) with stiff spines (Gabay et al. 1994), application of herbicide treatment causes problem with the contamination of the fruits (seeds) with toxins (Rady et al. 2014), and the efficiency of seed germination and seedling growth is low and is highly dependent on various biological and environmental factors (Abbasi et al. 2010). Therefore, in search for alternatives to production of desirable medicinal compounds from plants, cell culture technologies were introduced as a possible tool for studying and producing plant secondary metabolites (Vanisree et al. 2004). Biotechnological approaches by means of tissue cultures, gene transfer, molecular biology, and enzymology and bioreactor technology could be an alternative way for the production of flavonolignans.

Fig. 2.1 *Silybum marianum* plant

This chapter encompasses the recent publication on in vitro culture of *S. marianum*. Optimization of types of explant used, growth regulators rates, and the culture environment was employed to establishment of cell, callus, organ, transgenic cultures. However, abiotic and biotic elicitors and precursor feeding to enhancement accumulation of silymarin in different cultures of *S. marianum* are highlighted.

2 Biosynthesis of Silymarin in *S. marianum*

Little information has been published about the silymarin biosynthesis and its regulation (Dewick 2002). Silymarin is a collection of flavonoids and composed of taxifolin and at least seven flavonolignans, such as silybin A, silybin B, isosilybin A, isosilybin B, silychristin, isosilychristin, and silydianin (Biedermann et al. 2014), which can be extracted from the medicinal herb milk thistle. Among which, silybin is the major flavonolignan. However, the biosynthetic pathway of silybin remains unclear (Lv et al. 2017a). In this section, the biosynthetic pathway of silymarin in *Silybum marianum* and characterization of the related genes encoding the enzymes involved in this pathway are summarized in Fig. 2.2.

Producing the flavanone naringenin (Ng)
Silymarin is synthesized through the phenylpropanoid pathway trans-forming phenylalanine into 4-coumaroyl-CoA, which enters both the flavonoid and the monolignol pathway. Chalcone synthase(CHS) and chalcone isomerase (CHI) are the enzymes involved in the two-step condensation of three molecules of malonyl-CoA and one of 4-coumaroyl-CoA, producing the flavanone naringenin (Ng).

Formation of taxifolin (Tx)
Flavanone naringenin (Ng) is oxidized by flavanone 3-hydroxylase (F3H) to yield dihydrokaempferol (DHK). This latter compound is hydroxylatedon the 3_position

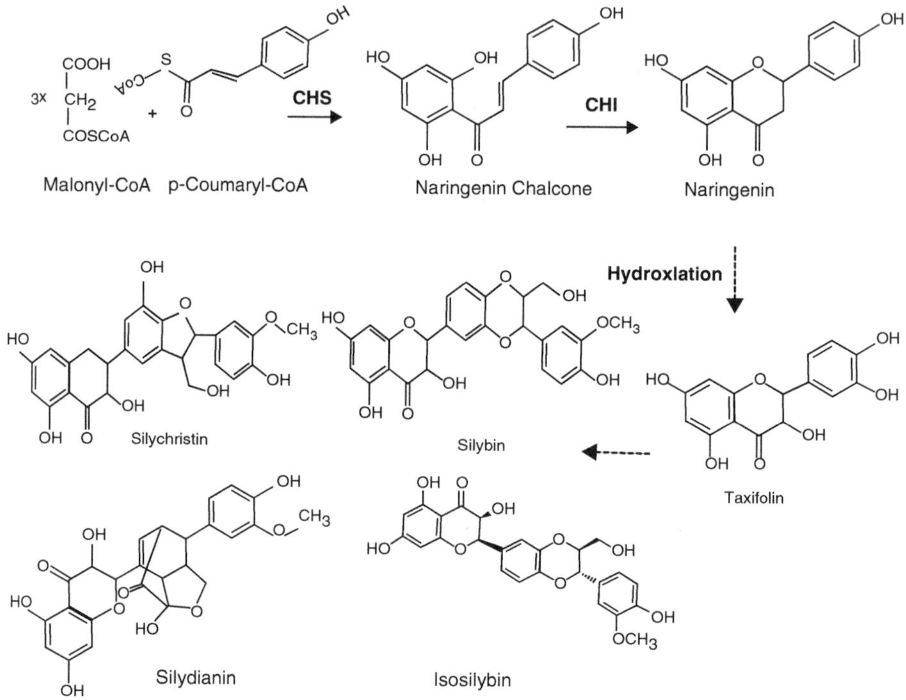

Fig. 2.2 Suggested pathway for the biosynthesis of silymarin components. CHS, chalcone synthase; CHI, chalcone isomerase. (Adapted from Rahnama et al. 2013)

of the B-ring by flavonoid 3-hydroxylase (F3H) giving the dihydroflavonol taxifolin (Tx), which is the precursor of the flavonoid part of Sm (Petrussa et al. 2013). On the other hand, coniferyl alcohol (CA) conforms the lignan moiety of the Sm molecule. Apart from the enzymes that give rise to 4-coumaroyl-CoA, seven other enzymes are involved in monolignol biosynthesis; among them, (hydroxy) cinnamyl alcohol dehydrogenase (CAD) is responsible for the formation of CA (Boerjan et al. 2003).

Silymarin formation
Silymarin formation occurs via oxidative radicalization of Tx and CA, followed by combinatorial radical coupling. Multiple orientation possibilities for coupling of the CA moiety to Tx give rise to the regioisomers of the Sm mixture: silychristin, isosilychristin, silydianin, silybin A, silybin B, isosilybin A, and isosilybin B (Morazzoni and Bombardelli 1995; Kim et al. 2003; Lee and Liu 2003).

Silymarin is a collection of compounds extracted from the medicinal herb milk thistle, among which silybin is the major flavonolignan. However, the biosynthetic pathway of silybin remains unclear. Biomimetic reactions demonstrated that silybin can be synthesized from coniferyl alcohol and taxifolin by the action of peroxidase. The concentration profiles of silybin and its precursors and RNA-Seq analysis of gene expression revealed that the amount of taxifolin and the activity of peroxidase serve as the limiting factors in silybin biosynthesis. Hierarchical clustering of the expression profile of genes of the flavonoid biosynthetic pathway distinguished flowers from other organs. RNA-Seq revealed five candidates for the peroxidase involved in silybin production, among which APX1 (ascorbate peroxidase 1) showed a distinct peroxidase activity and the capacity to synthesize silybin. The spatial organization of silybin biosynthesis in milk thistle was elucidated, which could help our understanding of the biosynthesis of silybin and other flavonolignans (Lv et al. 2017a).

3 In Vitro Culture of *S. marianum*

3.1 Cell and Callus Cultures of S. marianum

Callus tissue means it is an unorganized proliferate mass of cells. Callus can be initiated in vitro by placing small pieces of the whole plant (explants) onto a growth supporting medium under sterile conditions under the stimulus of endogenous growth substances or regulating chemicals added into the medium. The metabolism of cells which were in a quiescent state is changed, and they begin active division. During this process, cell differentiation and speculation, which may be occurring in the intact plant, are reversed, and the explants give rise to a new tissue which is composed of meristematic and unspecialized cell types (John and Koperuncholan 2012). Due to problems in traditional cultivation and extraction of its bioactive compound (silymarin), different strategies have been employed in order to fulfil the scarce demand of *S. marianum.* In vitro initiation of cell and callus cultures from different explants of *S. marianum* has been attempted by many workers. Studies on cell and callus cultures induction of *S. marianum* are presented in Table 2.1.

Leaf slices of *S. marianum* were inoculated on SH medium supplemented with 0.05 mg/l BA and 0.5 mg/l 2,4-D and solidified with 0.8% agar. Specific growth rate reached a maximum when the SH medium was supplemented with 0.05 mg/l 2,4-D and 0.06 mg/l kinetin. Cell suspension cultures were obtained using friable calli that were transferred to SH liquid medium supplemented with 0.05 mg/l BA and 0.5 mg/l 2,4-D. The results showed that growth inhibition was observed at sucrose concentrations higher than 3% (Feveriro et al. 1986). Also, Liu and Cai (1990) induced calli from hypocotyl explants of *S. marianum* when cultured on MS medium with 0.8 mg/l NAA and 0.5 mg/l BA. Iqbal and Srivastava (2000) obtained callus cultures from leaf, shoot apex, and nodal segments of in vivo and in vitro seedlings of *S. marianum.* The explants callused within 4 weeks on MS medium supplemented with NAA (0.1 mg/l) + BAP (0.3 mg/l) + zeatin (0.3 mg/l). Optimization of callus cultures of *S. marainum* for biomass production as a potential source of milk clotting peptidases was investigated (Cimino et al. 2006). They reported that B5 medium supplemented with 0.05 mg/l of BA and 0.5 mg/l of 2,4-D and solidified with 2.5 g/l of phytagel was the best for callus production from cotyledon explants of in vivo seedlings. Manaf et al. (2009) found that hypocotyl

Table 2.1 Studies on cell and callus cultures induction of *S. marianum*

Explant used	Medium and culture condition	Growth regulators (mg/l)									Reference
		2,4-D	NAA	Kin	BA	IAA	Ze	GA_3	TDZ	IBA	
Young leaves	SH-16/8 h photoperiod at 24 °C	0.5			0.05						Feveriro et al. (1986)
Hypocotyl	MS		0.8		0.5						Liu and Cai (1990)
Leaf, shoot apex, and nodal segments	MS—cultures were maintained under 10/14 h light/dark cycle in the culture room at 25 ± 2 °C		0.1		0.3		0.3				Iqbal and Srivastava (2000)
Cotyledons from in vivo seedlings	B5—cultures incubated at 24 °C in darkness during 45 days	0.5			0.05						Cimino et al. (2006)
Hypocotyl	MS—cultures were maintained at 25 ± 1 °C		1.0		0.1						Manaf et al. (2009)
Leaves from wild-grown plants	MS—cultures were maintained in a growth room at 25 ± 1 °C under 16 h photoperiod				5.0						Abbasi et al. (2010)
Hypocotyl- derived calli for cell suspension	MS—cultures were incubated at 25 °C in the dark	1.0		0.1							El-wekeel et al. (2012)
Leaves	MS—cultures were maintained in a growth room at 25 ± 1 °C under a 16 h photoperiod	2.5									John and Koperuncholan (2012)
Hypocotyl	MS	4.5									
Leaves	MS—cultures were maintained at 25 ± 2 °C and 16 h photoperiod			5.0		0.5					Bekheet et al. (2013)
Leaves	MS—cultures were maintained at 25 ± 2 °C and 16 h photoperiod		5.0	2.0				0.1	0.1		Bekheet et al. (2014)
Leaves	MS—cultures were incubated in the growth chamber at 25 ± 2 °C under a 16 h photoperiod	0.25		0.25							Rady et al. (2014)
leaf explants from 28-days old seedlings	MS—cultures were placed in 16 h photoperiod at 25 ± 1 °C								2.2 μM		Khan et al. (2014)
Petiole	SH—cultures were placed in 16 h photoperiod at 25 ± 1 °C	2.5			1.5						Khan et al. (2015a)
Leaves and roots	MS—maintained in a growth room at 25 ± 2 °C under 16 h photoperiod	0.5	0.5		0.5						Riasat et al. (2015)

Roots—for cell suspension Transfer to chapter—root cultures	MS liquid—continuous agitation at 125 rpm on orbital shaker at 25 °C in the dark		2.0	0.1							
Hypocotyl	MS—cultures were maintained in light condition for 16 h/day photoperiod at 25 ± 2 °C for 28 days		0.75		1.0						Manaf et al. (2016)
Cotyledons	MS—cultures were maintained at 25 ± 2 °C and under darkness			5.0		0.5					Gabr, et al. (2016)
Leaf explants	MS—cultures were maintained at 25 ± 2 °C under a 16 h photoperiod		5.0	2.0				0.1			
Cotyledons	MS—cultures were incubated in the growth chamber at 25 ± 2 °C in darkness	0.25			0.05						Elhaak, et al. (2016)
Petiole	B5—cultures were placed in 16 h photoperiod at 25 ± 1 °C	1.5							1.5		Abbasi et al. (2016)
Immature leaves (from wild plant)	MS—cultures were maintained at 25 ± 2 °C under a 16 h photoperiod		2.0		1.5						Lv et al. (2017a)
Root explant	MS—cultures were placed in a growth chamber at 16 h light/ 8 h dark photoperiod and 25 ± 2 °C	1.0		1.5							Eari et al. (2017)
Leaf explant		0.5		1.0							
Hypocotyl	MS—cultures were incubated at 25 ± 1 °C and 16 h light and 8 h dark photoperiod	1.0			0.5						Elsharnouby and Hassan (2018)
Leaf (28 days old in vitro derived Plantlets)	MS—cultures were incubated with a temperature 25 ± 1 °C and 16/8 h light/dark cycle		1.0		2.5						Younas et al. (2018)
Hypocotyl explant from 20-day-old seedling	MS—cultures were incubated under 16 h light and 8 h dark photoperiod at 25 ± 2 °C	1.0				1.0				1.0	Jabeen et al. (2019)
Leaf explant		1.0			0.5						

MS Murashige and Skoog (1962), *SH* Schenk and Hildebrandt (1972), *B5* Gamborg et al. (1968)

explant from *S. marianum* was the best explant for callus formation. MS medium supplemented with NAA (1.0 mg/l) + BA (0.1 mg/l) showed the highest callus fresh weight in light condition 16 h/day.

Abbasi et al. (2010) found that the highest frequency of callus induction was observed on leaf explants incubated on MS medium supplemented with 5.0 mg/l BA after 20 days of culture. Also, El-wekeel et al. (2012) established callus cultures from hypocotyl segments of *S. marianum*, when cultured on MS solid medium supplemented with 1 mg/l 2,4-D and 0.1 mg/l kinetin. Cell suspension cultures were established from 3-month-old callus grown on MS liquid with the same growth regulators in the dark on a rotary shaker at 100 rpm. The results of a study by John and Koperuncholan (2012) showed that MS medium with 2,4-D at 2.5 mg/l was the best for initiation of callus from leaf explants of *S. marianum* seedlings, while the medium contained 2,4-D at 4.5 mg/l resulted in 100% callus formation from hypocotyl explants. Also, Bekheet et al. (2013) found that the highest frequency of callus induction was observed on leaf explants cultured on MS medium supplemented with 5 mg/l kin +0.5 mg/l IAA. Callus growth parameters revealed that medium containing 5 mg/l kin +0.5 mg/l IAA + 0.1 mg/l GA_3 gave the highest fresh weight and growth value.

Bekheet et al. (2014) obtained callus cultures from leaf explants grown on MS medium supplemented with 5 mg/l NAA + 2 mg/l Kin +0.1 mg/l GA3. The results showed that addition of 3 mg/l picloram or 75 μM salicylic acid or 50 μM jasmonic acid to the culture medium generally enhanced callus growth values. Rady et al. (2014) obtained callus cultures from leaf explants of different *S. marianum* genotypes when cultured on MS medium supplemented with 0.25 mg/l 2,4-D and 0.25 mg/l kinetin. However, callus cultures were induced from leaf explants excised from in vitro seedlings of *S. marianum* cultured on MS medium supplemented with 2.2 μM TDZ; percentage callus induction was 73 (Khan et al. 2014). In another study, Khan et al. (2015a) reported that the callus formation frequency was significantly enhanced when the petiole explants from in vitro germinated seeds of *S. marianum* were cultured on SH medium with 2.5 mg/l 2,4-D in combination with 1.5 mg/l BA. In another study, Riasat et al. (2015) reported that green, friable, and rhizogenic callus was observed when small segments of *S. marianum* leaf and root explants were incubated on the MS medium containing 0.5 mg/l each of BAP, NAA, and 2,4-D together under light conditions. The results showed that the leaf explants gave better results in the callus initiation than root explants.

In recent studies, Manaf et al. (2016) reported that hypocotyl explant from *S. marianum* gains the best response for callus formation (callus growth index %, 95.12) on MS medium supplemented with 1.0 mg/l BAP + 0.75 mg/l NAA. Gabr et al. (2016) found that cotyledon explants of *S. marianum* cultured on MS medium containing 5.0 mg/l Kin and 0.5 mg/l IAA resulted in callus initiation, while leaf explants cultured on MS medium supplemented with 5.0 mg/l NAA + 2.0 mg/l Kin +0.1 mg/l GA_3 induced callus tissues. The results of morphological appearance of callus showed that addition of 200 mg/l chitosan to the culture medium has induced a protocorm-like body (PLB) in almost all callus, while at 400 mg/l chitosan, the PLB were compact and later differentiate into few long roots. The treatment of

10 mg/l of salicylic acid showed dark brown callus differentiated from each side several long primary roots. At 20 mg/l of MeJA, very few roots were formed from the dark brown callus. Elhaak et al. (2016) found that the greatest callus growth of cotyledon explants excised from seeds of *S. marianum* was observed on MS medium supplemented with 0.25 mg/l 2,4-D and 0.05 mg/l BAP, 50 mg/l asparagines, and 50 mg/l inositol. The suitable growth regulator concentrations combination of medium resulted in 4.6 and 2.5 times elevated callus FW and DW. Callus water content was strongly influenced by media composition in a time-dependent manner. Also, high 2,4-D level in the medium with intermediate concentrations of asparagines, BAP, and inositol resulted in maximum volume of callus after 21 days of cultivation.

Callus formation of *S. marianum* was induced from all explant types tested on B5 medium while MS or SH was only effective in leaf and petiole explants. Highest callus induction frequency (90%) was recorded when petiole explants were cultured on B5 medium supplemented with 1.5 mg/l 2,4-D and 1.5 mg/l thidiazuron (TDZ) (Abbasi et al. 2016). It was found that the highest frequency (98.9%) of callus induction from leaf explants of wild-grown *S. marianum* was obtained after 4 weeks on MS medium supplemented with 1.5 mg/l BA and 2.0 mg/l NAA (Lv et al. 2017b). Moreover, Eari et al. (2017) reported that, increasing 2,4-D concentration along with Kin, the callus induction has been increased significantly in root explants of *S. marianum*. Induced callus was friable and white in color. The highest percentage (98) of callus induction was observed in root explants grown on MS medium supplemented with 1.0 mg/l 2,4-D + 1.5 mg/l Kin, whereas maximum callus induction percentage (77) in leaf explants was observed on MS medium supplemented with 0.5 mg/l 2,4-D + 1.0 mg/l Kin. The induced callus was whitish yellow and friable. The highest fresh (12.33 g) and dry weight (0.870 g) were recorded in leaf-derived callus.

Elsharnouby and Hassan (2018) obtained big creamy friable callus cultures from hypocotyl segments of *S. marianum* grown on MS medium supplemented with 1 mg/l 2,4-D and 0.5 mg/l of BA. Both concentration and period of copper sulfate elicitation significantly affected the fresh and dry weight of callus. The medium supplemented with 5 μM copper sulfate after 28 days resulted in the highest growth index of callus (13.2%). Younas et al. (2018) found that combination of BAP (2.5 mg/l) and NAA (1.0 mg/l) resulted in brown friable leaf-derived callus of *S. marianum*. With optimum callogenic response (97%) when placed under cool white light with 16 h light and 8 h dark. Therefore, they suggested that white light is the best for callus induction. More recently, Jabeen et al. (2019) revealed no significant difference in days to callogenesis in the different explants (leaf, stem, and hypocotyl excised from *S. marianum* seedling). Highest callus induction frequency (100%) was recorded in hypocotyl when the MS medium was fortified with combination of 1.0 mg/l 2,4-D + 1.0 mg/l IBA + 1.0 mg/l IAA. The morphology of callus was soft, whitish yellow, granular, and highly proliferated. However, the leaf explant showed 95% callus induction frequency when cultured on MS + 1.0 mg/l 2,4-D + 0.5 mg/l BAP. Nature of the callus was greenish yellow, hard, compact, friable, and well proliferated. In addition, stem explant was not responsive to all employed medium for callus induction.

3.2 Plant Regeneration of S. marianum

Different research work has been achieved to explore protocols for in vitro regeneration of plants from *S. marianum.* Studies on in vitro regeneration of *S. marianum* are presented in Table 2.2.

Hypocotyl calli were induced on MS medium with 0.8 mg/l NAA and 0.5 mg/l BA. Shoots were regenerated from the surface of calli after transferring to D6 medium. The frequency of shoot differentiation was 75%. Whole plants with healthy root were obtained on MS medium containing 0.5 mg/l NAA and 0.1 mg/l IBA (Liu and Cai 1990). Indirect organogenesis was occurred with leaf, shoot apex, and nodal segment explants of *S. marianum* when grown on MS medium with NAA (0.1 mg/l), BAP (0.3 mg/l), and zeatin (0.3 mg/l). In these medium, callus tissues were formed, and upon transfer to MS medium containing 0.1 mg/l NAA and 0.5 mg/l zeatin, the callus differentiated multiple shoots followed by rooting. Direct organogenesis was induced when nodal explants were grown on MS medium with 0.1 mg/l IAA and 0.5 mg/l Kin. The obtained shoots were rooted on MS medium with 0.1 mg/l NAA and 0.5 mg/l zeatin (Iqbal and Srivastava 2000). Direct regeneration was achieved by cotyledons node explant of *S. marianum* cultured on MS medium supplemented with BA (3.0 mg/l) which gave the best shoot formation (20 shoot/explant). Root induction was observed when shoots were grown on MS medium with NAA (1.0 mg/l) (Manaf et al. 2009). Shoot organogenesis of *S. marianum* was also induced from leaf explants of wild-grown plants (Abbasi et al. 2010). Transfer of callogenic leaf explants onto MS medium supplemented with 2.0 mg/l GA_3 and 1.0 mg/l NAA resulted in 25.5 shoots per culture flask after 30 days of cultivation. Moreover, upon transferred shoots to elongation medium, the longest shoots were observed on MS medium supplemented with 0.5 mg/l BA and 1.0 mg/l NAA. The obtained shoots were rooted on MS free of plant growth regulators (Fig. 2.3). John and Koperuncholan (2012) found that MS medium supplemented with NAA at 2 mg/l and BAP at 1.5 mg/l was the best medium for initiation of shoots from hypocotyl-derived callus of *S. marianum.* Upon transferring the shoot cultures to MS medium with 2 mg/l GA_3, shoots were elongated. The MS medium containing 2 mg/l NAA was the best for rooting of the produced shoots. Bekheet et al. (2013) obtained complete plantlets through direct organogenesis when leaf explants of *S. marianum* were cultured on MS medium with 1 mg/l BA and 2 mg/l NAA. Also, indirect organogenesis was occurred when leaf-derived callus was grown on MS medium with 2 mg/l Kin and 2 mg/l NAA. In another study, multiple shoots and leaves were induced from shoot tip explants of in vitro grown *S. marianum* seedlings when grown on MS medium with 2.0 mg/l BA. Maximum number (25.6) of shoots per explants was recorded when shoots were transferred to MS medium containing 1.0 mg/l BA with 0.1 mg/l NAA. The produced shoots were rooted on MS medium with 0.2 mg/l IAA. Maximum number of roots as well as highest root length was 11.0 and 2.4 cm, respectively (El-Sherif et al. 2013).

Shoot cultures were established from shoot tip explants from aseptically germinated seeds of *S. marianum*. The best medium for initiation of the highest number

Table 2.2 Studies on in vitro regeneration of *S. marianum*

Explant type	Establishment stage	Multiplication stage	Rooting stage	Reference
	Response			
Hypocotyl	MS; NAA (0.8 mg/l) + BA (0.5 mg/l)	D6 medium	MS; NAA (0.5 mg/l) + IBA (0.1 mg/l)	Liu and Cai (1990)
Leaf and shoot apex	MS; NAA (0.1 mg/l) + BA (0.3 mg/l) + Ze (0.3 mg/l)	MS; NAA (0.1 mg/l) + Ze (0.5 mg/l)		Iqbal and Srivastava (2000)
Nodal segments	MS; IAA (0.1 mg/l) + Kin (0.5 mg/l)		MS; NAA (0.1 mg/l) + Ze (0.5 mg/l)	
Cotyledons node	MS; BA (3.0 mg/l)		NAA (1.0 mg/l)	Manaf et al. (2009)
Leaves from wild-grown plants	MS; BA (5.0 mg/l) MS; NAA (1.0 mg/l) + GA_3 (2.0 mg/l)	MS; NAA (1.0 mg/l) + BA (0.5 mg/l) (Elongation of shoots)	MS; basal (Free hormone)	Abbasi et al. (2010)
Hypocotyl callus	MS; NAA (2.0 mg/l) + BA (1.5 mg/l)	MS; GA_3 (2.0 mg/l)	MS; NAA (2.0 mg/l)	John and Koperuncholan (2012)
Leaves	MS; NAA (2.0 mg/l) + BA (1.0 mg/l)			Bekheet et al. (2013)
Leaf-derived callus	MS; NAA (2.0 mg/l) + kin (2.0 mg/l)			
Shoot tips	MS; BA (2.0 mg/l)	MS; NAA (0.1 mg/l) + BA (1.0 mg/l)	MS; IAA (2.0 mg/l)	El-Sherif et al. (2013)
Shoot tips	MS; NAA (0.25 mg/l) + BA (0.25 mg/l)			Rady et al. (2014)
Nodal segments	MS; NAA (0.1 mg/l) + Kin (0.5 mg/l)	MS; Kin (1.6 mg/l)	MS; NAA (1.0 mg/l)	Al-Hawamdeh et al. (2014)
Leaf-derived callus	MS; Kin (6.6 μM) + TDZ (4.4 μM)	MS; NAA (4.4 μM) + TDZ (6.6 μM)	MS; IAA (0.4 μM)	Khan et al. (2014)
Petiole-derived calli	SH; 2,4-D (1.5 mg/l) + BA (1.5 mg/l) (Initiation of somatic embryos)	½ MS basal (Somatic embryos maturation)	MS; GA_3 (1.5 mg/l) (Germination and plant conversion)	Khan et al. (2015a)
Leaves	MS; NAA (1.0 mg/l) + IBA (1.0 mg/l)			Riasat et al. (2015)

(continued)

Table 2.2 (continued)

Explant type	Establishment stage	Multiplication stage	Rooting stage	Reference
	Response			
Petiole-derived callus	B5; 2,4-D (1.5 mg/l) + TDZ (1.5 mg/l) (Embryogenic callus formation)	B5; 2,4-D (0.5 mg/l) + TDZ (1.5 mg/l) Initiation of somatic embryos)	MS; basal (growth and maturation of somatic embryos)	Abbasi et al. (2016)
Leaf-derived callus	MS; GA_3 (3.0 mg/l) + NAA (1.0 mg/l)	MS; NAA (0.3 mg/l) + BA (1.0 mg/l)	½ MS; IBA (1.0 mg/l)	Lv et al. (2017a)

TDZ Thidiazuron, *D6 medium* = MS + NAA (0.8 mg/l) + BA (2.0 mg/l)

Fig. 2.3 Plant regeneration from leaf explants of *Silybum marianum*. (**a**) Callus formation from leaf explant, Bar 5 mm. (**b**) Shoot organogenesis from leaf explants, Bar 0.7 cm. (**c**) Proliferation and elongation of regenerated shoots, Bar 1.2 cm. (**d**) Rooting of regenerated shoots, Bar 2.5 cm. (Adapted from Abbasi et al. 2010)

of shoots (5) was MS medium with 0.25 mg/l each of BA and NAA, while medium without hormones gave the tallest (11.33 cm) shoots (Rady et al. 2014). Also, shoot initiation was occurred when nodal explants of *S. marianum* were grown on MS medium with 0.5 mg/l Kin and 0.1 mg/l NAA. Multiple shoots were regenerated on MS medium with 1.6 mg/l Kin. The highest number of shoot and shoot length were 5.40 and 6.72 cm, respectively. The obtained shoots were rooted on MS medium with 1.0 mg/l NAA. The highest rooting percentage, number of root, and root length were 72, 3.6, and 3.04 cm, respectively. Rooted plantlets were successfully transferred to plastic cups containing 1 peat:1 perlite, and the survival percentage was 70%. Acclimatized plants appeared normal and did not exhibit any morphological abnormalities (Al-Hawamdeh et al. 2014). In another study, Khan et al. (2014) found that leaf-derived callus of *S. marianum* initiated on MS medium with 2.2 μM TDZ (thidiazuron) gave the highest callogenic response (89%) when subcultured on MS medium with 4.4 μM (TDZ) in combination with 6.6 μM Kin. Subsequent subculturing of callus, on medium with similar compositions of PGRs, induced shoot organogenesis. Highest shoot induction frequency (86%) with maximum mean multiple shoots (26 shoots per explant) were recorded for MS medium with 11 μM TDZ. Longest shoots (4.1 cm) were recorded with 6.6 μM TDZ and

4.4 μM NAA. However, for shoot elongation the medium with 6.6 μM TDZ and 4.4 μM NAA gave the longest mean shoot length (4.1 cm). Furthermore, rooted plantlets were developed on MS medium with 0.4 μM IAA. The highest rooting frequency was 87.32% with mean number of roots (3.2 roots) and mean root length (3.4 cm). The rooted shoots were transferred to sterilized vermiculite and kept under growth room conditions for 2 weeks prior transfer to pots. The survival rate of rooted shoots was 73%.

Highest embryogenic potential was observed for calli previously derived from petiole explants of *S. marianum* cultured on SH medium containing 2.5 mg/l 2,4-D and 1.5 mg/l BA (Khan et al. 2015a). They found that highest number of somatic embryos (46 somatic embryo per callus) was observed on SH medium with 1.5 mg/l 2,4-D and 1.5 mg/l BA. However, half-strength MS medium showed optimal response for maturation of embryos (60%). For germination of embryos and plantlet conversion, highest germination rate (70%) was observed on MS medium with at 1.5 mg/l GA_3. Plantlets were then transplanted to pots with a mixture of soil, sand, and perlite (1,2:1, v/v/v) for acclimatization.

Induction of white compact callus, shoot elongation, and rooting were observed when leaf explants from in vitro germinated seeds of *S. marianum* were cultured on MS medium containing 1 mg/l NAA. Direct regeneration and rooting were induced when leaf explants were grown on MS medium with 1 mg/l IBA (Riasat et al. 2015). Recently, Abbasi et al. (2016) found that highest embryogenic callus formation (77%) was observed with petiole explant-derived callus of *S. marianum* cultured on B5 medium + 1.5 mg/l 2,4-D + 1.5 mg/l thidiazuron (TDZ). When the embryogenic calluses were subcultured onto B5 medium containing 0.5 mg/l 2,4-D + 1.5 mg/l TDZ, 77% of the cultures responded with 39.1 somatic embryos per callus. Maximum numbers of somatic embryos (31 somatic embryos per callus) were inspected at 2,4-D (1.0 mg/l) alone from petiole-derived embryogenic callus. Furthermore, MS basal medium was indicated more responsive for growth and maturation of somatic embryos. More recently, Lv et al. (2017a) reported that the optimal medium for shoot organogenesis from leaf explants of wild-grown *S. marianum* was MS supplemented with 3.0 mg/l GA_3 and 1.0 mg/l NAA, while shoot proliferation was accomplished in MS supplemented with 1.0 mg/l BA and 0.3 mg/l NAA. Rooting (94.5%) of these shoots was achieved after 4 weeks on 1/2 MS medium supplemented with 1.0 mg/l IBA. A total of 92% of the plants were surviving 6 weeks after transplantation of plantlets to soil. RAPD analysis revealed that the genetic fidelity between the parent and the in vitro raised plantlets exhibited 100% similarity.

3.3 *Root Cultures of* S. marianum

Biotechnological approaches have been established for the production of secondary metabolites through root cultures where the undifferentiated cultures (callus and cell suspension) are not able to produce these compounds efficiently as compared to the root cultures (Bais et al. 2001). Adventitious root culture system provides an

Table 2.3 Studies on initiation of root cultures of *S. marianum*

Explant used	Medium used	Culture condition	Growth regulators (mg/l) NAA	IBA	Kin	Reference
Leaflets	MS	Cultures were maintained in a growth room at 25 ± 1 °C under a 16 h photoperiod	2.0		0.2	John and Koperuncholan, (2012)
Young shoots	MS		0.1	0.1		Iri et al. (2013)
Roots obtained from in vitro grown seedlings	MS (Liquid)	Cultures were kept under continuous agitation at 125 rpm on orbital shaker at 25 °C in the dark		2.0		Riasat et al. (2015)
Nodal explants from 4-week-old in vitro germinated seedlings	MS (Liquid)	Cultures were incubated under 24 h dark with a continuous shaking in gyratory shaker at 110 rpm at room temperature	1.0			Khan et al. (2015b)

alternative approach for the improvement and enhancement of plant-based pharmaceutical compounds. It serves as a base for conservation of important plant species by reducing pressure on its natural populations (Ahmad et al. 2015). Adventitious root formation is an intricate process involving various endogenous and exogenous factors. According to Han et al. (2001), "the process can be divided into three stages: root induction (includes molecular and biochemical changes occurring before any cytological event); root initiation (first anatomical modifications) and root protrusion (emergence of root primordial)." These roots show high stability and growth rate and synthesize appropriate amount of secondary metabolites in intercellular spaces which can be easily extracted and can be grown in a phytohormone amended medium with low inoculums (Sivakumar et al. 2006). Hairy root cultures have attractive attention for production of secondary metabolites from medicinal plants as it is stable, grow faster, and have the same or greater biosynthetic capacity to produce the secondary metabolites compared to plant cell cultures and mother plants (Kim et al. 2002). But the transformed hairy root cultures produce opine-like substances which are lethal to mammalian cells and not always accepted (Yoshikawa and Furuya 1987). All these observations have oriented workers to use organized plant tissue culture which has grown continuously such as root cultures for the production of secondary metabolites.

The present section focuses on the initiation and establishment of adventitious root cultures from *S. marianum* through various explants in vitro. Studies on initiation of root cultures of *S. marianum* are shown in Table 2.3.

Direct root regeneration was carried out using different concentrations of auxin and cytokinin such as NAA and KIN. The best results were observed in the range of NAA at 2.0 mg/l and Kin at 0.2 mg/l, after 20 days of cultivation (John and Koperuncholan 2012). However, Iri et al. (2013) indicated that the highest root formation percentage and root length were observed in young shoots of *S. marianum* grown in MS medium supplemented with 0.1 mg/l each of IBA and NAA. Riasat et al. (2015) obtained adventitious root cultures of *S. marianum* from root explants of in vitro grown plantlets cultured on MS liquid medium supplemented with 2 mg/l IBA. Recently, Khan et al. 2015b) induced adventitious roots from nodal explants of *S. marianum* grown on solid MS medium supplemented with 1.0 mg/l of NAA. Four-week-old fresh adventitious roots were transferred to MS liquid medium. In between 8 weeks of culture period, the highest fresh biomass formation was observed in 6-week-old cultures. This finding was reported by Cui et al. (2010) who found that IBA at concentration 1 mg/l was suitable for the adventitious root induction in *H. perforatum*.

3.4 Genetic Transformation of S. marianum

3.4.1 *Agrobacterium tumefaciens*-Mediated Transformation

Several investigators have carried out biotechnological efforts at in vitro level to elucidate the biosynthetic pathway and its genes to enhance the production of these compounds by genetic modulation of plant. Reports in the literature concerning *A. tumefaciens*-mediated transformation of *S. marianum* are scarce. However, there is only one available recent report on genetic transformation of *S. marianum*. This study is presented in Table 2.4.

The growing demand for *t*-resveratrol for industrial uses has generated considerable interest in its production. Heterologous resveratrol production in plant cell suspensions, apart from requiring the introduction of only one or two genes, has the advantage of high biomass yield and a short cultivation time and thus could be an option for large-scale production (Hidalgo et al. 2017). The potential of metabolically engineered *S. marianum* cultures for *t*-resveratrol production was

Table 2.4 Studies on *A. tumefaciens*-mediated transformation of *S. marianum*

Infected explant/ culture	*Agrobacterium* strain/ gene construct	Results/response	Reference
Suspension cultures	Strain C58C1 (pGV2260) harboring a binary vector pJCV52-STS3 with VvSTS3 gene construct	The stilbene synthase (STS) gene was present in all the transgenic clones, but not in the wild type Integration and expression of the VvSTS3 gene in the *S. marianum* genome make cell suspension cultures competent for *t*-R production and release to the culture medium under elicitation conditions A significant amount of *t*-R was obtained from the extracellular medium of 30 mM methylated *β*-cyclodextrins (Cd)-treated cultures, increasing from the beginning of elicitation up to about 12 mg/l at 76 h	Hidalgo et al. (2017)

studied (Hidalgo et al. 2017). Cell suspensions were stably transformed with *Vitis vinifera* stilbene synthase 3, and the expression of the transgene led to extracellular *t*-resveratrol accumulation at the level of milligrams per liter under elicitation. Resveratrol synthesis occurred at the expense of coniferyl alcohol. Percentage of increase of coniferyl alcohol (CA) and flavanolignan production both in non-transformed and STS-transformed cultures 24 and 48 h after methylated *β*-cyclodextrins (Cd) treatment. From the present results, they concluded that S. *marianum* cell cultures represent a new heterologous host for *t*-R production.

3.4.2 *Agrobacterium rhizogenesis*-Mediated Transformation

Hairy roots are genetically and biochemically stable, and their ability to biosynthesis of secondary metabolites is comparable to the original plant cells. These, as well as their fast growth rate, have made them attractive for production of secondary metabolites (Rahnama et al. 2008). Studies in establishment of hairy root cultures from *S. marianum* have been attempted by few workers as shown in Table 2.5.

Hairy roots from *S. marianum* were induced by inoculation of explants with *A. rhizogenes* (Rahnama et al. 2008). All explants were precultured 3 days on MS hormone-free medium and then immersed on overnight cultures of bacterial suspension for 10 min. One-month-old hairy roots were dissected from the explants and grown in (MS) liquid medium. The results showed that transformation efficiencies were 7.9% for hypocotyls, 21.6% for cotyledons, and 20% for whole plants by using the injection method. PCR analysis confirmed the presence of the GUS and *rol*B gene. Rahnama et al. (2013) used petunia chalcone synthase (chsA) gene to enhance the silymarin production level in the hairy root cultures of *S. marianum*. Hypocotyl explants were precultured 3 days on MS hormone-free medium and then inoculated with an overnight suspension culture of *A. rhizogenes* for 10 min. The explants were transferred to the selection medium, the co-culture medium supplemented with 250 mg/l cefotaxime and 50 mg/l kanamycin. Bacteria-free hairy roots were transferred to hormone-free liquid MS medium and incubated in the dark at 25 °C on a rotary (Fig. 2.4). Molecular analysis confirmed the presence and integration of chsA transgene in transgenic hairy roots. Chemical analysis indicated that the overexpression of chsA gene enhanced the silymarin production level in the transgenic line as much as sevenfold than the non-transgenic hairy roots. Bekheet et al. (2013) established hairy root cultures from cotyledon explant of *S. marianum*. The explants were immersed in bacterial suspension for 10 min and incubated in the dark at 28 °C. After 24 h of co-cultivation, the explant tissues were transferred to new growth medium (solidified free MS medium) containing 500 mg/l cefotaxime to eliminate bacteria and then incubated in growth chamber at 25 ± 2 °C in dark condition. It was found that numerous hairy roots were observed emerging from the wound sites within 4–5 weeks. Hairy roots were formed in high frequency on wounded regions of the young

Table 2.5 Hairy root culture studies on *S.marianum*

Infected explant	*Agrobacterium* strain	Response	Reference
Hypocotyl, leaf, cotyledons, and whole plant	AR15834 strain harboring the binary vector pBI121 which carries genes *nptll* coding for neomycin phosphotransferase II driven by the nopaline synthase promoter and an *uidA* gene coding for β glucoronidase (Gus) driven by (CaMV) 35S promoter Optical density at 600 nm, (OD_{600} = 0.7)	From 150 cotyledon explants inoculated with *A. rhizogenese* containing the pBI121 vector, 48 roots were produced after 4 weeks PCR analysis confirmed the presence of the GUS and *rol*B genes Histochemical GUS staining of root tissues confirmed GUS activity in 45 (30%) of the hairy root clones	Rahnama et al. (2008)
Hypocotyl	AR15834 strain harboring the pCamCHS vectors AR15834 strain free from external plasmids (for the induction of the control hairy root)	PCR analysis confirmed the presence of chsA, rolB, and nptII genes in three lines of transgenic hairy roots, while non-transgenic hairy roots only carried the rolB gene RT-PCR analysis confirmed the expression of chsA gene at transcript level in the transgenic hairy roots The integration of the chsA transgene into the genome of transgenic hairy roots was confirmed by Southern blot analysis	Rahnama et al. (2013)
Cotyledons (12 days old)	A4 strain	Hairy roots were induced within 4 7weeks after inoculation Using young leaf explants resulted in enhancing transformation frequency	Bekheet et al. (2013)
Different explants	Six strains were used (R1601, R15384, R1000, A4, R1025, and R1)	Integration of the DNA plasmids in the *A. rhizogenes* into the genome of transformed roots was confirmed by PCR Using liquid chromatography, it is determined that the silibin content in hairy roots was 2.5 times that in the plant	Zhang et al. (2014)

(3-week-old) leaves. However, PCR analysis confirmed the presence of *rolA* and *rolC* genes in transformed hairy roots. However, Zhang et al. (2014) established hairy roots from different explants from *S. marianum* using different strains of *A. rhizogenes***.** The results showed that all of the six *A. rhizogenes* used generated hairy roots. The strain A4 showed comparatively high infection on the explants. PCR confirmed the transfer of DNA plasmids in the *A. rhizogenes* into the genome of the transformed roots.

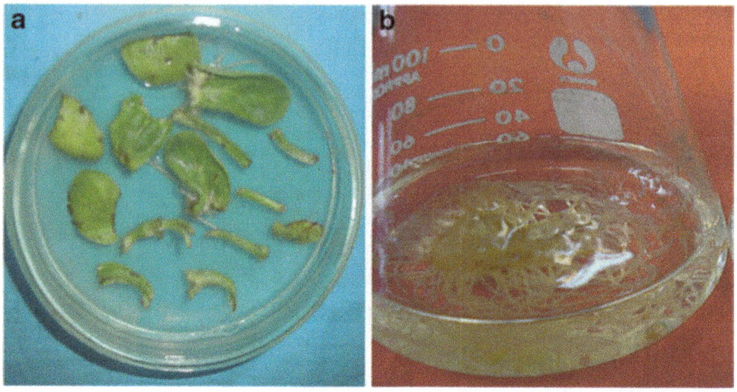

Fig. 2.4 Hairy root induction in *S. marianum*. (**a**) Hairy root induced on explants. (**b**) Hairy root culture in MS liquid medium. (Adapted from Rahnama et al. 2013)

4 Silymarin Production from In Vitro Culture of *S. marianum* Through Biotic and Abiotic Elicitation and Precursor Feeding

4.1 Abiotic Elicitation and Silymarin Production from In Vitro Cultures of Silybum marianum

An elicitor is a compound which can simulate a stress condition in the culture, triggering the same response in the cells that would be induced in the natural environment, usually resulting in the accumulation of secondary metabolites (Zhao et al. 2005; Vasconsuelo and Boland 2007). Abiotic elicitation means triggering the synthesis of phytochemicals in plants by using chemical or physical stimuli (Owolabi et al. 2018).

4.1.1 Genotype and Type of Explants

Distinct chemotypic variations in *S. marianum* have been discovered in different ecotypes/cultivars and genotypes (Shokrpour et al. 2007; Poppe and Petersen 2016). These variations in genotypes caused changes in relative amounts of silymarin components from different origins and reflected in in vitro culture. It was reported that the silymarin content in fruits depends on the milk thistle variety and geographic and climatic conditions in which they grow. Response of in vitro cultures of *S. marianum* to genotype and type of explants is presented in Table 2.6.

Table 2.6 Response of in vitro cultures of *S. marianum* to genotype and type of explants

Culture type	Abiotic elicitor	Response	Reference
Callus cultures from cotyledon and hypocotyl	Two genotypes (ecotype) (Budakalszi and Noor abad moghan cultivars)	The best cultivar for silymarin production was Budakalszi cultivar	Pourjabar et al. (2012)
Suspension cultures derived from cotyledons callus tissues under shaking (120 rpm) in the dark at 26 °C	Three typical chemotypes (BGMr-Marburg, Klenk-Poland, and BGF-Frankfurt)	The flavonolignan content extracted from cell suspension cultures is much lower compared to the amounts detectable in fruit coats of milk thistle There are only minute amounts of flavonolignans present in the medium	Poppe and Petersen (2016)
Cultures from in vitro explants	Explants (leaf, shoot apex, and nodal segments)	Nodal segment calli yielded maximum silybin	Iqbal and Srivastava (2000)
		Regenerants from leaf explants showed the highest yield of silybin	
		Nodal segment calli showed the highest yield of silybin	
Cultures from in vivo explants	Explants (leaf, shoot apex and nodal segments)	All the in vitro explant-derived calli and regenerants produced higher amounts of silybin than those from in vivo	
Callus cultures from cotyledon and hypocotyl	Explants (cotyledon and hypocotyl)	Calli produced from cotyledon explant of Budakalszi cultivar were superior for silybin A + B, where calli from hypocotyl explants produced higher amount of taxifolin and silydianin + silychristin	Pourjabar et al. (2012)
Leaf and root explants and their derived callus	Explants (leaves and roots)	Leaf explant is the most responsive to produce higher fresh weight Leaf and root explants showed similar total silymarin amount	Eari et al. (2017)

4.1.1.1 Genotypes

Pourjabar et al. (2012) obtained callus tissues from two explants (cotyledon and hypocotyls) of two genotypes (Budakalszi and Noor abad moghan cultivars) of *S. marianum*. Result of analysis of variance showed that flavonolignan content of calli was not significant difference between the two genotypes. The best cultivar for silymarin production was Budakalszi cultivar. Also, Poppe and Petersen (2016) obtained suspension cultures from cotyledon callus of three typical chemotypes of *S. marianum*. The in vitro cultivated cells of the three different chemotypes already revealed visible differences in their phenotype. The results showed that dried suspension cells only accumulate about 1–10% of the amount of silymarin extracted from fruits. The total silymarin amount itself varied between 0.01 and 0.02% of the cell dry wt on average, decreasing slowly with higher numbers of subcultures. Silymarin components released to the cell culture medium showed a highly differing composition with only low amounts of silychristin and silydianin. However, in vitro cultures showed a similar pattern in the cells as in whole fruits but not in the cell culture medium.

4.1.1.2 Explants

Iqbal and Srivastava (2000) established callus cultures from different explants of *S. marianum* from in vivo plants and in vitro seedlings. Of all the calli produced from in vivo and in vitro explants, nodal segment callus yielded maximum silybin content. In general, regenerants from in vitro explants showed higher amounts of silybin than those from in vivo explants. Pourjabar et al. (2012) found that calli produced from *S. marianum* cotyledon explant of Budakalszi cultivar were superior for silybin A + B, where calli from hypocotyl explant produced higher amount of taxifolin and silydianin+silychristin. Recently, Eari et al. (2017) revealed that the amount of silymarin in root and leaf explants from *S. marianum* seedlings was 2.86 and 3.94 mg/g dw, respectively. They concluded that leaf explant-derived callus was more responsive to growth and silymarin content.

4.1.2 Medium Composition, pH, and Subculture Times

Different research work has been achieved to explore the effect of nutrient medium composition, pH, and subculture times on the accumulation of silymarin. Response of in vitro cultures to medium composition, pH, and subculture times is presented in Table 2.7.

4.1.2.1 Nutrient Medium Composition

One of the strategies most commonly employed to improve secondary metabolite production in in vitro cultures is manipulation of the components of the culture medium. Silymarin accumulation was not enhanced either upon using different concentrations of KNO_3, KH_2PO_4, or iron or by changes in the levels of sugars in the nutrient medium. However, application of nutrient stress through the elimination of calcium ions did reduce growth and promoted flavonolignan production (Cacho et al. 1999). It was also suggested that inhibition of external and internal calcium fluxes plays a significant role in flavonolignan metabolism of *S. marianum* cell cultures. Elimination of calcium ions from the medium of cell suspension cultures increased flavonolignan production (Sanchez-Sampedro et al. 2005a). They also reported that silymarin accumulation was not altered by treatment of cultures with the calcium ionophore A23187. The specific Ca2+ chelator, EGTA, enhanced the

Table 2.7 Response of medium composition, pH, and subculture times

Culture type	Abiotic elicitor	Response	Reference
Cell suspension (from 3-month-old hypocotyl callus)	Calcium-deprived medium	Elimination of calcium ions from the medium increased flavonolignan production	(Sanchez-Sampedro et al. (2005a)
	[a]EGTA (0.1, 1.0, and 3 mM culture age 6 days).	A strong release of silymarin into the culture medium was observed at 3 mM in complete calcium medium	
	Verapamil or $LaCl_3$ (0.01, 0.1 and 1.0 mM)	0.1 mM of both compounds was the minimal level tested for which silymarin production was increased in calcium-containing medium	
	Calcium ionophore A23187 (20 μM) 24 h treatment	Prolonged incubation times, or 20 μM level, reduced the accumulation of silymarin	
	Ruthenium red, thapsigargin, and [b]TMB-8 (0.01, 0.1 and 1.0 mM) after 72 h of treatment	The three compounds induced silymarin secretion into the complete culture medium	
Cell suspension culture from callus initiated from hypocotyl explants	MS medium with different concentrations of medium major components	0.5 phosphate concentration from $KH_2PO_4 + K_2SO_4$, 0.75 nitrogen concentration, 0.75 phosphate concentration from $KH_2PO_4 + K_2SO_4$, glucose 30 g/l, and $CaCl_2$ half concentration recorded the highest silymarin content to reach 57.5%, 46%, 24.2%, 21.2%, and 19.6%, respectively, over the control	Rabie et al. (2010)
Shoot cultures (from seedlings)	Carbon sources (glucose, fructose, and sucrose in conc. 15, 30, 45, 60 mg/l)	Glucose at (45 g/l) gave the highest silymarin content	Al-Hawamdeh et al. (2013)
In vivo plants		In vivo leaves samples showed higher % of silymarin yield than in vitro samples	
Root cultures	Different volumes of medium culture	Did not show any significant difference on root growth and silymarin production	Iri et al. (2013)
Root cultures	Different pH medium	Culture medium with pH = 7 was superior to other pH for silymarin production	Iri et al. (2013)
Hairy roots (from 30-day-old cultures)	pH (5, 5.7, 6 and 7) under 25 /25 °C for 30 days	Maximum silymarin concentration (0.25 mg/g DW) was obtained in cultures treated with pH 5, whereas the silymarin accumulation was 0.18, 0.15, and 0.18 mg/g DW in pH 5.7, 6, and 7, respectively	Rahimi and Hasanloo (2016)

(continued)

Table 2.7 (continued)

Culture type	Abiotic elicitor	Response	Reference
Hairy roots (from 30-day-old cultures)	pH (5, 5.7, 6, and 7)	Lipoxygenase (LOX) activity was strongly affected by pH. Acidic environment inducing signal for LOX activity and increased silymarin production	
Callus cultures from cotyledon and hypocotyl	Subculture times	The effect of subcultures on the amount of TXF, SBN A + B, and also ISBN A + B had significant difference, but the amount of SDN + SCN had not significant difference	Pourjabar et al. (2012)

EGTA, a specific calcium chelator able to reduce the availability of extracellular calcium
[a]EGTA, ethylene glycol-bis(b-aminoethyl) ether-N, N, N0, N0 tetraacetic acid
[b]TMB-8, 8-(diethylamino)octyl 3, 4, 5-trimethoxybenzoate

silymarin content in cells by 200% and its secretion by 3–4 times. However, the inorganic ion La3+, as well as the calcium channel inhibitor verapamil, also stimulated production. Several reagents known to block intracellular calcium movement, such as ruthenium red, thapsigargin, and TMB-8, appreciably increased silymarin accumulation. Also, Rabie et al. (2010) showed that the presence of KH_2PO_4 (0.085 g/l) + K_2SO_4 (0.05445 g/l) in the suspension culture medium of hypocotyl-derived callus from *S. marianum* gave the highest callus fresh, dry weights and growth index, 7.26 0.488 g and 13.53, respectively. MS medium with different concentrations of medium major components resulted in increasing silymarin content in cells over the control.

Nutrient medium composition or selection of medium also played a vital role for elicitation process. Glucose was found to be a better carbon source than fructose and sucrose. Silymarin accumulation (1.63%) was observed more in shoot cultures of *S. marianum* when glucose at 45 g/l was added to the MS medium as compared to other carbon sources, fructose, and sucrose which accumulated at 45 g/l amount of 1.42 and 1.40% of total silymarin, respectively (Al-Hawamdeh et al. 2013). In another study, Iri et al. (2013) established shoot cultures from hypocotyl explants of *S. marianum*. Roots initiated from shoots were cultured in liquid medium and subjected to different volumes of culture medium. They found no significant difference on silymarin production in root cultures.

4.1.2.2 Different pH Medium

Iri et al. (2013) found that root cultures of *S. marianum* grown on culture medium (pH = 7) were superior to other pH for silymarin production. Recently, Rahimi and Hasanloo (2016) reported that maximal increases in biomass and silymarin accumulation occurred in the root cultures of *S. marianum* grown in pH 5. Dry biomass was 0.50 g compared to pH 5.7, 6, and 7 (0.24, 0.27, and 0.21 g), respectively.

In cultures treated with pH 5, maximum silymarin content was 0.25 mg/g DW, whereas the silymarin accumulation was 0.18, 0.15, and 0.18 mg/g DW in cultures treated with pH 5.7, 6, and 7, respectively.

4.1.2.3 Subculture Times

Also, age of subculture or subculture times plays an important role in production of bioactive compounds by elicitation. In a study by Pourjabar et al. (2012), it was found that subculture times of *S. marianum* callus cultures had significant difference in accumulation of TXF, SBN A + B, and also ISBN A + B. Amount of SBN A + B was increased from 2.718% at first calli to 6.015% at the fourth subculture, while amount of TXF and ISBN A + B was decreased from 0.205 and 0.122% at first calli to 0.12 and 0.00% at the fourth subculture, respectively. On the other hand, amount of SDN + SCN had no significant difference.

4.1.3 Light and Dark, Gamma Irradiation, and Temperature

Environmental stresses result in an increased formation of toxic active oxygen species which may cause breakdown of membranes and lipid peroxidation (Mittler et al. 2004). Light is an important factor for both ex vitro and in vitro morphogenetic study. Its presence, absence, time, and intensity directly influence anabolic and catabolic processes, particularly secondary metabolism (Seibert and Kadkade 1980). Also, light is considered as a key abiotic elicitor influencing several physiological processes in plants, including the biosynthesis of secondary metabolites (Younas et al. 2018). Importance of light or dark for increasing production of secondary metabolites in cell cultures of S. marianum was investigated by several workers and shown in Table 2.8.

4.1.3.1 Light and Dark

Light-grown cell suspension of *S. marianum* showed reduced growth in terms of dry weight. The higher contents of silymarin were recorded in cell cultures grown in the dark compared to that of cultures grown in the light in media containing 3 or 2 mg/l picloram and 2 mg/l jasmonic acid (Hasanloo et al. 2008a). In contrast, Iri et al. (2013) found that light treatment of root cultures induced more silymarin compared to dark treatment. Recently, Younas et al. (2018) reported that, in leaf-derived callus cultures of *S. marianum*, red light significantly increased the total phenolic content (TPC), total flavonoid content (TFC), antioxidant and superoxide dismutase (SOD) activities while highest peroxidase (POD) activity for the dark-grown cultures, followed by green light-grown cultures. HPLC analysis revealed enhanced total silymarin content under red light (18.67 mg/g DW), which was almost double than

Table 2.8 Response of in vitro cultures of *S. marianum* to light and dark, gamma irradiation, and temperature

Culture type	Abiotic elicitor	Response	Reference
Cell cultures derived from cotyledon explants	Light and dark	The light-grown cell suspension cultures showed reduced growth in terms of dry weight and silymarin content	Hasanloo et al. (2008a)
Root cultures	Light and dark	Light treatment induced more silymarin compared to dark treatment	Iri et al. (2013)
Leaf-derived Callus cultures	Several spectres of monochromatic lights (blue LEDs (460 nm, 24 h); green LEDs (510 nm, 24 h); red LEDs (660 nm, 24-h); yellow LEDs (570 nm, 24 h); white LEDs (400–700 nm, 24 h); darkness (24 h); cool white fluorescent tubes (400–700 nm, 16/8 h, photoperiod, as control)).	HPLC analysis revealed enhanced total silymarin content under red light Green spectrum resulted in highest accumulation of isosilybin A and isosilybin B Amount of taxifolin was found maximum under continuous white light	Younas et al. (2018)
In vitro shoot cultures and in vivo plants	Gamma irradiation (0.0, 5,10 and 15 Gy)	The highest total silymarin conc. was found in in vitro regenerated plants when exposed to (10 Gy) gamma irradiation dose	El-Sherif et al. (2013)
Hairy roots (from 30-day-old cultures)	Temperature (30 °C/25 °C, 25 °C/25 °C, and 15 °C/20 °C in 16 h/8 h cycle)	25 °C/25 °C and acidic environment of medium are beneficial for silymarin production	Rahimi and Hasanloo (2016)

control (9.17 mg/g DW), whereas green spectrum resulted in highest accumulation of isosilybin A and isosilybin B. Conversely, the amount of taxifolin was found maximum under continuous white light (0.480 mg/g DW) which was almost eight-fold greater than control (0.063 mg/g DW).

4.1.3.2 Gamma Irradiation

Gamma rays belong to ionizing radiation and interact on atoms or molecules to produce free radicals in cells. These radicals can damage or modify important components of plant cells and have been reported to affect differentially the morphology, anatomy, biochemistry, and physiology of plants depending on the irradiation level (Kim et al. 2004; Wi et al. 2005; El Sherif et al. 2011). The highest total silymarin level was found to be 6.598% dry weight in the in vitro regenerated shoot tip explants when exposed to (10 Gy) gamma irradiation dose. However, the number of leaves and plant height decreased with increasing gamma irradiations (El Sherif et al. 2013).

4.1.3.3 Temperature

The effect of temperature on root biomass and silymarin production in hairy root cultures of *S. marianum* was evaluated (Rahimi and Hasanloo 2016). They reported that incubation temperature, 25 °C/25 °C, promoted root biomass, and the silymarin content was 0.18 mg/g DW in 4-week-old hairy roots as compared with the cultures treated with 15 °C/20 °C and 30 °C/25 °C (0.13 and 0.12 mg/g DW, respectively).

4.1.4 Growth Regulators

Plant growth regulators (PGRs) have been used as proficient elicitors to stimulate production of plant secondary metabolites. Traditionally, synthetic plant growth regulators are used as precious research tools to elucidate physiological responses of plants or to explore biochemical control mechanisms (Jamwal et al. 2018). Response of in vitro cultures of *S. marianum* to growth regulators is shown in Table 2.9.

Iqbal and Srivastava (2000) reported that nodal segment-derived callus originated from in vivo or in vitro plants of *S. marianum* and maintained on MS medium supplemented with NAA (0.1 mg/l) and zeatin (0.5 mg/l) gave the best silybin content, 0.90 and 1.30 mg/g dry weight, respectively. However, leaf of regenerated plants yielded the highest content of silybin (1.90 mg/g dry weight). Hasanloo et al. (2008a) found that MS medium supplemented with 3 mg l/l picloram and 2 mg/l jasmonic acid gave the highest dry weight (32 mg/g DW) and silymarin content (0.413 mg/g DW) in cell suspension cultures of *S. marianum*. Also, Hasanloo et al. (2009) found that MS medium supplemented with 0.1 mg/l NAA gave the highest growth rates (3.04 g) and silymarin content (0.15 mg/g DW, threefold higher than control) in hairy root cultures of *S. marianum*. Pourjabar et al. (2012) found that callus cultures of *S. marianum* grown on MS medium supplemented with 2, 4-D (2 mg/l) and Kin (1 mg/l) gave the highest silybin A + B content (5.59%) than other treatments. Al-Hawamdeh et al. (2013) found that the presence of Kin or BA or 2ip in concentration of 1.6, 2, and 1 mg/l in the culture medium of in vitro grown shoot cultures of *S. marianum* gave the highest total silymarin, 1.33, 1.04, and 1%, respectively, than other treatment and control. Otherwise, the in vivo (wild) grown shoots gave 1.53% total silymarin. Elhaak et al. (2016) found that MS medium supplemented with 0.25 mg/l 2,4-D, 0.05 mg/l BAP, 50 mg/l asparagine, and 50 mg/l inositol gave the highest callus mass production and the greatest amount of silymarin in callus cultures of *S. marianum*, where 118 mg dry weight callus tissues produced 2.659 mg/g dry weight silymarin in comparison with other treatment. Moreover, the content of silymarin in seeds was lower than produced from callus grown in all the used media. However, Manaf et al. (2016) found that the highest value of silymarin concentration (65.85 μg/g) was noticed when callus cultures of *S. marianum* were

Table 2.9 Response of in vitro cultures of *S. marianum* to growth regulators

Culture type	Abiotic elicitor (Growth regulators)	Response	Reference
Cultures from in vivo and in vitro explants	Different growth regulators	Cultures maintained on MS + NAA (0.1 mg/l) + zeatin (0.5 mg/l) yielded best silybin content	Iqbal and Srivastava (2000)
Cell cultures from cotyledon explant	Picloram, jasmonic acid (in dark)	The greatest growth rates and silymarin content were obtained with 3 mg l/l picloram and 2 mg/l jasmonic acid in darkness after 28 days	Hasanloo et al. (2008a)
Hairy roots (30-day-old)	Effects of auxins	The highest growth rates and silymarin content (threefold higher than control) were observed by addition of 0.1 mg/l NAA to the culture medium	Hasanloo et al. (2009)
Callus cultures from cotyledon and hypocotyl	Growth regulators Kin (0.1, 1.0 or 2.0 mg/l) and 2,4-D (1.0 or 2.0 mg/l) in dark	The maximum amount of silybin A + B was obtained from medium with 2,4-D (2.0 mg/l) and Kin (1 mg/l)	Pourjabar et al. (2012)
Shoot cultures (from seedlings)	Growth regulators (kin, BA, and 2ip in conc. 0.0, 0.4, 1.0, 1.6, or 2.0 mg/l)	MS; Kin (1.6 mg/l) + NAA (0.1 mg/l) gave the highest silymarin content compared with control and other treatments	Al-Hawamdeh, et al. (2013)
Callus cultures derived from cotyledon explant	Different growth regulators	MS; 2,4-D (0.25 mg/l) + BAP (0.05 mg/l) + asparagines (50 mg/l) + inositol (50 mg/l) gave the highest callus mass production and the greatest amount of silymarin	Elhaak et al. (2016)
Hypocotyl-derived callus	Different growth regulators (BAP and NAA)	MS; BA (1.25 mg/l) + NAA (0.57 mg/l) was the best for silymarin accumulation in callus tissues	Manaf et al. (2016)
Leaf- and root-derived callus	Different concentrations of 2,4-D and Kin	Maximum silymarin percent was observed in root-derived callus on MS medium supplemented with 0.5 mg/l 2,4-D and 1.0 mg/l Kin	Eari et al. (2017)

grown on MS mediun with BA (1.25 mg/l) + NAA (0.57 mg/l). Recently, Eari et al. (2017) revealed that combination of 0.5 mg/l 2,4-D and 1.0 mg/l Kin gave the highest amount of silymarin amount, 33.88 and 8.76 mg/g dw in root- and leaf-derived callus of *S. marianum*, respectively.

4.1.5 Methyl Jasmonate

Methyl jasmonate (MeJA) is one of the jasmonic acid derivatives, which strongly promoted the accumulation of silymarin. MeJA is a ubiquitous plant signaling messenger involved in diverse regulatory functions such as defense response against wounding, pathogens, temperature, and salinity stress (Sivanandhan et al. 2013).

The exogenous application of methyl jasmonate (MeJA) has been carried out to enhance the biosynthesis of silymarin in different cultures of *S. marianum*. Response of in vitro cultures of *S. marianum* to methyl jasmonate is presented in Table 2.10.

Cell suspension from hypocotyl-derived callus of *S. marianum* was supplied at day 3 with different MeJA concentrations. Silymarin (silychristin + silybinins) was analyzed in the cells and medium after 72 h of treatment. The optimum silymarin concentration was found to be at 100 μM MeJA. At this concentration, silymarin content were 1.98 mg/g DW for cells and 0.088 mg/50 ml for the culture medium (Sanchez-Sampedro et al. 2005b). Addition of MeJA in undifferentiated hypocotyl callus cultures of *S. marianum* caused a metabolic reprogramming that affected amino acid and carbohydrate metabolism; upon elicitation sucrose decreased and glucose levels increased (Sanchez-Sampedro et al. 2007). In cell suspension cultures derived from hypocotyl callus, a concentration of 100 μM

Table 2.10 Response of in vitro cultures of *S. marianum* to methyl jasmonate

Culture type	Abiotic elicitor (Methyl jasmonate)	Response	Reference
Cell suspension (hypocotyl-derived callus)	Methyl jasmonate (MeJA) 100 μM	MeJA was effective in eliciting silymarin in cells and medium	Sanchez-Sampedro et al. (2005b)
Cell suspension (from 3-month-old hypocotyl callus)	Methyl jasmonate	Methyl jasmonate caused increasing in glucose levels and decreasing in sucrose level in cultures	Sanchez-Sampedro, et al. (2007)
Cell suspension (hypocotyl-derived callus)	Methyl jasmonate (MeJA) 100 μM	MeJA increased the activity of the enzyme phospholipase D (PLD)	Madrid and Corchete (2010)
Cell cultures developed from hypocotyl-derived callus. Cell cultures were incubated in the dark at 25 °C and shaken at 90 rpm	Methyl jasmonate (100 μM)	Delayed addition of MJ to cultures subcultivated in medium containing RMCD markedly influenced Sm production by promoting conversion of the previously formed coniferyl alcohol precursor	Belchi-Navarro et al. (2011)
Callus tissues derived from cotyledon (leaves), hypocotyl	Methyl jasmonate (MeJA) (20, 40, and 80 mg/l)	MeJA (20 mg/l) showed the highest silymarin accumulation compared to other two elicitors (chitosan and salicylic acid)	Gabr et al. (2016)
Cell cultures were developed from cotyledon and hypocotyl explants from 3-month-old callus Cultures were incubated at 25 °C and shaken at 90 rpm in darkness	Methyl jasmonate (100 μM)	SLM content of the MeJA treated cell cultures increased significantly after 48 h No significant difference between the cell culture derived from cotyledon and hypocotyl in terms of silymarin content	Pourjabar et al. (2018)

MeJA increased the activity of the enzyme phospholipase D (PLD) which caused a substantial increase in silymarin production (Madrid and Corchete 2010).

Belchi-Navarro et al. (2011) reported that production of suspension cultures of *S. marianum* treated with methyl jasmonate can be increased by inclusion of cyclodextrins (CDs) in the cultures. Both hydroxylated (RHCD) and dimethylated (RMCD) CDs strongly induced prompt accumulation of coniferyl alcohol (CA) in the medium followed by a late production of flavonolignans. Simultaneous addition of methyl jasmonate (MJ) and RMCD to cells did not significantly modify CA release or flavonolignan accumulation. Delayed addition of MJ to cultures subcultivated in medium containing RMCD markedly influenced silymarin production by promoting conversion of the previously formed CA precursor. However, Gabr et al. (2016) found that addition of 20 mg/l MeJA in medium of callus cultures derived from cotyledon accumulated the highest silymarin content (13.59 μg/g dw) compared to control treatments (3.70 μg/g dw). Recently, Pourjabar et al. (2018) found that cell suspension derived from callus tissues of *S. marianum* and subjected to MeJA treatment (100 μM) resulted in increased amount of silymarin production to 1.23-fold in comparison to the control after 48 h. The highest amount of silymarin (approximately 1.5 mg/g dw) was obtained in cell suspension cultures derived from cotyledon treated with MeJA, than hypocotyl-derived callus.

4.1.6 Salicylic Acid

Salicylic acid has been described as an essential component in the activation of numerous plant defense responses and can be used as an elicitor for secondary metabolites production (Shah and Klessig 1999). Salicylic acid is considered to have a role in the regulation of plant growth and development and regulates resistance to pathogens like bacterial, fungal, and viral. It specifically binds to catalysis and inhibits the activity of the enzyme (Chen et al. 1993). Thus, exogenous salicylic acid can increase H_2O_2 content of tissue and induce the expression of antioxidant enzymes and increase plant tolerance to the abiotic stressor. Increase in silymarin concentration with respect to salicylic acid addition could be attributed to the fact that salicylic acid acts as a powerful inducing signal. For silymarin production, it is known to regulate the expression of various defense-related genes (Shualev et al. 1995). Similarly, salicylic acid acts as a vital stress signaling component widely studied for its involvement in systemic acquired resistance (SAR) against diseases and pathogens (Sivanandhan et al. 2013). Response of in vitro cultures of *S. marianum* to salicylic acid is presented in Table 2.11.

Khalili et al. (2009) reported that a reduction in dry weight of *S. marianum* hairy root cultures was observed by the addition of salicylic acid. The dry weight decreased to 0.82 g compared with non-treated hairy roots (1.11 g) after 72 h of supplementation with 8 mg/50 ml culture salicylic acid. Salicylic acid enhanced silymarin production (1.89 mg/g DW). The optimal feeding condition was the addition of salicylic acid (6 mg/50 ml culture) after 24 h in which the silymarin content was 2.42 times higher than the control (0.78 mg/g DW), while after 72 h treatment with salicylic acid (6 mg/50 ml culture). Silymarin accumulation was

Table 2.11 Response of in vitro cultures of *S. marianum* to salicylic acid

Culture type	Abiotic elicitor (Salicylic acid)	Response	Reference
Hairy root cultures (28 days old)	Salicylic acid (0, 1, 2, 4, 6, and 8 mg/50 ml culture) Exposure times (24, 48, 72, 96, and 120 h)	Addition of salicylic acid (6 mg/50 ml culture) caused increase in silymarin content 2.42 times higher than the control after 24 h	Khalili et al. (2009)
Hairy root cultures (28 days old)	Salicylic acid (6 mg/50 ml culture) for different times (24, 48, 72, 96, and 120 h)	The highest accumulation of H_2O_2 was 86.60 µmol/g DW that was obtained 120 h after elicitation (2 times of non-treated hairy roots) [a]IC_{50} increased upon treatment and was 1.15-times higher than the control 24 h after elicitation Tocopherol content under SA treatment was lower than the control	Khalili et al. (2010a)
		Maximum activity of peroxidase and ascorbate peroxidase were 0.17 and 1.50 µmol min-1 mg^{-1} protein, 96 and 72 h after elicitation, respectively	
Callus and cell suspension cultures	Salicylic acid (50 or 100 µg/ml) Added to the 7-day-old cultures for 48 h	A 1.6-fold increase in silychristin content was observed when using salicylic acid at conc. of 50 µg/ml salicylic acid-induced production of silybin that was not detected primarily in control	El-wekeel et al. (2012)
Callus tissues derived from cotyledon (leaves) Hypocotyl and hairy roots	Salicylic acid (SA) 10, 20, and 40 mg/l	Higher conc. of SA (20–40 mg/l) triggered accumulation of total silymarin	Gabr et al. (2016)

[a]IC_{50} mg DPPH/ml of extract

significantly increased from 0.47 mg/g DW in the control to 0.90 mg/g DW in hairy roots, (1.9-fold above the control). In addition, lipoxygenase activity and linoleic acid content of the hairy roots increased 24 h after elicitation with salicylic acid by 1.57- and 2.37-fold increase, respectively, than the control treatment. Also, Khalili et al. (2010a) reported that exogenous salicylic acid can increase H_2O_2 content in milk thistle hairy root cultures and induce the expression of antioxidant enzymes and increase secondary metabolite accumulation. El-wekeel et al. (2012) reported that addition of salicylic acid (50 µg/ml) in *S. marianum* cell suspension cultures gave the highest silychristin content (0.18 mg/g DW) compared to control (0.11 mg/g DW), after 48 h elicitation, while salicylic acid (100 µg/ml) gave the highest silybin content (0.04 mg/g DW) compared to control treatment (0.0 mg/g DW). Recently, Gabr et al. (2016) found that higher concentrations of salicylic acid (20 mg/l SA to 40 mg/l) triggered accumulation of total silymarin (9.63 and 11.24 ug/g DW), respectively, in undifferentiated hypocotyl callus of *S. marianum*.

4.1.7 Silver Nitrate ($AgNO_3$) and Copper Sulfate ($CuSO_4$)

Many authors reported that metals, such as Co^{2+}, Ni^{2+}, Fe^{2+}, Ag^+, and V, all stimulate the biosynthesis of a wide range of plant secondary metabolites (Tumova et al. 2001, Zhao et al. 2001). Ag^+ has been introduced as an inhibitor of the ethylene signal transduction pathway (Zhao et al. 2005). Response of in vitro cultures of *S. marianum* to silver nitrate and copper sulfate is presented in Table 2.12.

4.1.7.1 Silver Nitrate

The effectives of Ag + elicitation depend on elicitor concentrations and time of treatment. The timing of the hairy root harvesting is critical. It could be a result of switching from primary metabolism to secondary metabolisms. Besides these considerations, the positive effect of $AgNO_3$ on production, and specifically on silymarin accumulation, lipoxygenase activity, and linoleic acid content, would make this treatment an interesting candidate for improving productivity in large-scale processes (Khalili et al. 2010b).

Rahimi et al. (2010) reported that cell growth of *S. marianum* cell suspension was increased after 12 h in media supplemented with 0.2, 0.4, 0.8, and 1 mM Ag^+, and the value in medium treated by 1 mM Ag^+ after 72 h elicitor application was 7.21 g, which was about twofold that of the control (3.32 g). The highest silymarin production reached about 56 μg/g DW, 24 h after treatment with Ag^+ (0.8 mM), which was 30-fold, than that of the control. Khalili et al. (2010b) found that the elicited hairy root cultures of *S. marianum* gave the highest content of silymarin (0.563 mg/g DW) after 72 h elicitation, with 2 mM of silver nitrate which was two times of non-treated hairy roots (0.281 mg/g DW). In addition, in the next stage, hairy roots were treated with 2 mM silver nitrate for different exposure times (0, 24, 48, 72, 96, and 120 h). The highest silymarin production reached to 1.200 mg/g DW, 96 h after elicitation, which was three and half times of control treatment (0.326 mg/g DW). Vildová et al. (2014) found that the highest content of taxifolin production (2.2 mg/g DW) in cell suspension culture of *S. marianum* was detected after silver nitrate ($5.887.10^{-4}$ mol/l) treatment and 72 h application, while control samples do not produce secondary metabolites. However, in nutrient medium of cell suspension cultures, the highest content of taxifolin (1.9 mg/100 ml) and silychristin (1.7 mg/100 ml) was after the elicitor treatment (5.887×10^{-3} mol/l) and (5.887×10^{-4} mol/l) after 48 and 168 h application, respectively. The maximum contents of excreted silydianin (3.5 mg/100 ml), silybin A (1.8 mg/100 ml), and silybin B (1.2 mg/100 ml) were found after elicitor treatment (5.887×10^{-5} mol/l) and 12 h application. The control suspension culture released taxifolin (0.1 mg/100 ml) in 0 h and also silychristin (0.1 mg/100 ml) and silydianin (0.3 mg/100 ml) after 168 h into the nutrient medium.

Table 2.12 Response of in vitro cultures of *S. marianum* to silver nitrate ($AgNO_3$) and copper sulfate ($CuSO_4$)

Culture type	Abiotic elicitor	Response	Reference
Cell suspension derived from cotyledon calli (in the dark at 25 ± 1 °C)	Silver nitrate-Ag + (0.2, 0.4, 0.8, 1, and 2 mM) for different exposure times (12, 24, 48, 72, 144, and 216 h)	Increase in cell growth was observed after 12 h in media with 0.2, 0.4, 0.8, and 1 mM Ag^+	Rahimi et al. (2010)
		The highest silymarin production 24 h after treatment with Ag + (0.8 mM) was 30-fold that of the control	
Hairy roots (28-day-old cultures)	Silver nitrate $AgNO_3$ (0.2, 0.4, 0.8, 1, and 2 mM) and harvested 72 h after elicitation	The highest content of silymarin was obtained with 2 mM of Ag + which was 2 times of non-treated hairy roots	Khalili et al. (2010b)
Hairy roots	Silver nitrate $AgNO_3$ (2 mM) (exposure to different times (0, 24, 48, 72, 96, and 120 h)	The highest silymarin production reached to 96 h after elicitation. Ag + treated hairy roots grew more slowly than the non-treated hairy roots	
Cell suspension culture	Silver nitrate ($AgNO_3$) in various concentrations ($5.887.10^{-3}$ mol/l, $5.887.10^{-4}$ mol/l, and $5.887.10^{-5}$ mol/l) Samples were taken after 6, 12, 24, 48, 72, and168 hours of treatment)	The highest content of taxifolin was detected after silver nitrate (5.887.10–4 mol/l) treatment and 72 h application Silybin A, silybin B, silydianin, silychristin, isosilybin A, and isosilybin B were not produced	Vildová et al. (2014)
Nutrient medium		The maximum contents of excreted silydian, silybin A, and silybin B were found after elicitor treatment (5.887×10^{-5} mol/l) and 12 h application	
		The highest content of taxifolin was after the elicitor treatment (5.887×10^{-3} mol/l) and 48 h application	
		The highest content of released silychristin was after the elicitor treatment (5.887×10^{-4} mol/l) and 168 h application	
Callus cultures derived from hypocotyl explants (in the dark at 25 ± 1 °C)	Copper sulfate (CuSO4) was applied in concentrations of 0, 3, 5, 7, and 9 μM Elicitation periods used were 2, 4, 7, 14, and 28 days	A remarkable improvement in silymarin content was observed with use of 5 μM copper sulfate as an abiotic elicitor for 4 days as the increase was observed	Elsharnouby and Hassan (2018)

4.1.7.2 Copper Sulfate

Copper sulfate use, as an elicitor, for the improvement of secondary metabolites in various medicinal plants has been reported. Enhancement of flavonoid accumulation in cell cultures of *D. lanata* as a result of copper sulfate treatment was reported by Bota and Deliu (2011). They found that flavonoid production reached over 10 times more as compared to the control. In *S. marianum* callus cultures, Elsharnouby and Hassan (2018) found that silymarin was positively correlated with $CuSO_4$ levels since all levels of copper sulfate significantly enhanced its content in relative to the control. Additionally, more silymarin was accumulated after 4 or 7 days, and the accumulation significantly decreased when the elicitation period reached 14 days more. The highest silymarin content (11.79 and 11.67 mg/g DW) was obtained when 5 or 7 μM copper sulfate levels were combined with 4 days elicitation period, being about fivefold of the control.

4.1.8 NaCl, Mannitol, and Polyethylene Glycol

The concentrations of various secondary plant products strongly depend on the growing conditions, and it is obvious that stress situations have a strong impact on the metabolic pathways responsible for the accumulation of the related natural products (Nikolova and Ivancheva 2005). Response of in vitro cultures of *S. marianum* to NaCl, mannitol, and polyethylene glycol is presented in Table 2.13.

4.1.8.1 Sodium Chloride

El Sherif et al. (2013) found that shoots produced from explants of *S. marianum* grown on medium supplemented with different concentrations of sodium chloride (NaCl) showed increase in number of shoots/explants and shoot length and decrease in number of leaves/explants and explant fresh weight with increasing NaCl level. Doubling the NaCl concentration in the medium had a positive effect on total silymarin concentration of shoots which reached 1.808% dry weight. Recently, Manaf et al. (2016) found that callus tissues of *S. marianum* grown on medium containing NaCl (3000 mg/l) gave the highest significant values in silymarin concentration (58.12 μg/g) as compared with the rest treatments.

4.1.8.2 Mannitol

Bekheet (2015) found that supplementation of culture medium with mannitol slightly depressed both fresh weight and growth value of leaf-derived callus of *S. marianum*. However, mannitol caused increase of dry matter and silymarin content. It was also found that silybin B was more responsive to mannitol stress compared with the other component of silymarin. The highest values of silybin A (0.45 mg/g DW), silybin B (0.60 mg/g DW), silychristin (0.014 mg/g DW), and silydianin

Table 2.13 Response of in vitro cultures of *S. marianum* to NaCl, mannitol, and polyethylene glycol

Culture type	Abiotic elicitor	Response	Reference
In vitro shoot cultures and in vivo plants	NaCl (25 and 50 mM)	Doubling the NaCl conc. in the medium had a positive effect on total silymarin	El-Sherif et al. (2013)
Hypocotyl-derived callus	NaCl (0, 3000, 4000, and 5000 mg/l)	NaCl treatments inhibited the growth of callus NaCl (3000 mg/l) added to MS medium gave the highest values of silymarin	Manaf et al. (2016)
Leaf-derived callus	Mannitol (20, 40, and 60 mg/l)	Addition 40 mg/l of mannitol to culture medium was the best concentration for total silymarin production Treatment of callus culture with 40 g/l mannitol increased production of silybin B to a level about fivefold higher than that of the control	Bekheet (2015)
Leaf-derived callus	Polyethylene glycol (400) (PEG) 5, 10, and 15 g/l	Presence of PEG (15 g/l) in culture medium strongly enhanced silymarin accumulation	

(0.31 mg/g DW) were recorded with medium containing 40 g/l mannitol. Moreover silychristin component was less responsive to mannitol stress.

4.1.8.3 Polyethylene Glycol (PEG)

A progressive reduction in both fresh weight and growth value in leaf-derived callus of *S. marianum* treated with polyethylene glycol (PEG) was observed (Bekheet, 2015). The presence of PEG (15 g/l) increased both silybin A and B to a level more sevenfold higher than that of the control. However, the highest values of silybin A (0.73 mg/g DW), silybin B (0.85 mg/g DW), silychristin (0.047 mg/g DW), and silydianin (0.51 mg/g DW) were recorded when 15 g/l of PEG was added to culture medium.

4.1.9 Chemicals

Response of in vitro cultures of *S. marianum* to some chemicals is presented in Table 2.14.

4.1.9.1 Mastoparan (Mst) and Butanol

Madrid and Corchete (2010) found that addition of mastoparan to cell suspension cultures of *S. marianum* stimulated phospholipase D (PLD) activity and strongly enhanced silymarin production. Treatment with the same concentration of the inactive analogue mastoparan 17 was ineffective. They also added that treating the cell

Table 2.14 Response of in vitro cultures of *S. marianum* to some chemicals

Culture type	Abiotic elicitor	Response	Reference
Cell suspension (hypocotyl-derived callus)	Mastoparan (Mst) (a PLD activity stimulator)	Enhanced phospholipase D (PLD) and caused a substantial increase in silymarin production Phosphatidic acid (PA) promoted silymarin accumulation	Madrid and Corchete (2010)
	Iso-, sec-, or tert-butanol	No effect on silymarin production	
In vitro shoot cultures	Quercetin (0.5 and 1.0 mg/l)	No significant effect on silymarin accumulation	El-Sherif et al. (2013)
Callus cultures derived from cotyledon explant	Glutathione (0, 0.1, 0.5, 1.0, and 2.0 mM) Callus was harvested after long- and short-term period	Elicitation with glutathione at 0.1 mM after 8 days increased silymarin accumulation by 6.4% (short-term period)	Elhaak et al. (2016)
Callus cultures derived from cotyledon explant	Ascorbate (0, 0.1, 0.5, 1.0, and 2.0 mM) Callus was harvested after long and short-term period	Elicitation with ascorbate at 0.5 mM after 8 days increased silymarin accumulation by 30.6% (short-term period)	Elhaak et al. (2016)
Callus cultures from hypocotyl explants (12 h, 24 h, 48 h, 72 h, and 168 h after elicitor application) controls were run for 24 h and 168 h	5-brom-2 hydroxyphenyl) amide of 5-*tert*-butyl-6-chlorpyrazine-2-carboxylic acid at conc. of 2.59 × 104 mol/l (*c1*), 2.59 × 10^{-5} mol/l (*c2*), 2.59 × 10^{-6} mol/l (*c3*), and 2.59 × 10–7 mol/l (*c4*)	The flavonolignan accumulation after elicitation with 5-brom-2-hydroxyphenylamide of 5-*tert*-butyl-6 chloropyrazine-2-carboxylic acid (conc. of 2.59 × 10^{-4} mol/l) 24 h elicitation was increased by about 1039%	Tumova et al. (2005)
Cell suspension cultures (12 h, 24 h, 48 h, 72 h, and 168 h after elicitor application) controls were run for 24 h and 168 h	3-methylamide of 5-*tert*-butylpyrazine-2-carboxylic acid at conc. of 3.71 × 10^{-4} mol/l (*c1*), 3.71 × 10^{-5} mol·l-1 (*c2*), 3.71 × 10^{-6} mol/l (*c3*), 3.71 × 10^{-7} mol/l (*c4*)	The elicitation with 3-methylamide 5-*tert*-butylpyrazine −2-carboxylic acid at conc. of 3.71 × 10^{-7} mol/l and within 72 h of elicitation, caused increasing in flavonolignan production by 893% compared with control	
Callus cultures 6, 12, 24, 48, 72, and 168 h after elicitor application	*N*-(3-iodo-4-methylphenyl) pyrazine-2-carboxamide (1) c1 = 100 mg/100 ml (2.95·10–3 mol/l) c1a = 10 mg/100 ml (2.95·10^{-4} mol/l) c1b = 1 mg/100 ml (2.95·10–5 mol/l)	The greatest increase of flavonolignan and taxifoline production was observed at conc. c1a after 6, 24, and 72 h at conc. c1b. Increased production of silychristin was achieved after 6-h elicitation with c1a, it was 2 times higher compared to the control sample	Tůmová et al. (2010)

(continued)

Table 2.14 (continued)

Culture type	Abiotic elicitor	Response	Reference
	N-(3-iodo-4 methylphenyl)-5-*tert*-butyl pyrazine-2-carboxamide (2), c2 = 100 mg/100 ml. (2.53·10–3 mol/l) c2a = 10 mg/100 ml. ($2.53 \cdot 10^{-4}$ mol/l) c2b = 1 mg/100 ml ($2.53 \cdot 10^{-5}$ mol/l)	An increased production of silychristin was reached with compound 2 at the conc. c2 after 72 h of elicitation. The production of silychristin in this case was increased 12 times compared to control	
Callus cultures (12, 24, 48, 72, and 168 h) after elicitor application	5-(2 hydroxybenzoyl)-pyrazine-2-carboxamide) at conc. of 1.159×10^{-3} mol/l (c1); 1.159×10^{-4} mol/l (c2); 1.159 × 10–5 mol/l (c3)	The maximum content of silymarin complex was induced by 168 h elicitor application of a conc. 1.159×10^{-4} mol/l	Tumova et al. (2011)
Cell suspension (12, 24, 48, 72, and 168 h) after elicitor application		The maximum content of silydianin was induced by 24 h elicitor application in conc. of 1.159×10^{-3} mol/l	

suspension with n-butanol (nBuOH) 20 min prior to the addition of methyljasmonate (MeJA) or mastoparan, the alcohol, at 0.2% concentration, diminished silymarin accumulation in elicited cultures. Treatment of cultures with the alcohol alone also impeded silymarin secretion into the culture medium.

4.1.9.2 Quercetin

Quercetin contains a polyphenolic chemical substructure that stops oxidation by acting as a scavenger of free radicals that are responsible for oxidative chain reactions and can be used as a general antioxidant stress protectant for plants. Quercetin has been reported to inhibit the oxidation of other molecules and hence is classified as an antioxidant (Russo et al. 2014). Increasing quercetin concentrations in the culture medium caused increasing in number of shoots/explants and shoot length but had a negative effect on the number of leaves/explants and explant fresh weight. In addition, there was no significant effect on silymarin accumulation (El-Sherif et al. 2013).

4.1.9.3 Glutathione

Glutathione has been considered as a hub in cell redox metabolism due to its multiple and central functions. In this respect, Dubreuil-Maurizi et al. (2011) studied glutathione deficiency of the *Arabidopsis* mutant pad2–1 and stated that modifications in cell glutathione contents can alter gene expression and metabolism and induce expression of defense genes. Elhaak et al. (2016) reported that *S. marianum* callus growth rate (3.2%) was higher at 0.1 mM glutathione than other treatment but was lower than control (4%). After 42 days of cultivation, glutathione at concentration 0.1 mM in the medium increased accumulation of silymarin (21 mg/g dry weight) compared with control treatment (19 mg/g dry weight) and reduced it at higher concentrations.

4.1.9.4 Ascorbate

Ascorbate is directly involved in the regulation of two processes that mediate morphogenic responses in plants: cell division and elongation in addition to the regulation of mitotic activity in the meristems. The effect of different concentrations of ascorbate on *S. marianum* callus growth and silymarin production after 42 days of cultivation was investigated by Elhaak et al. (2016). They found that ascorbate had a positive effect on callus growth rate with increasing the concentration up to 1 mM. Silymarin content was 26 mg/g dry weight at 2 mM ascorbate and equal to control treatment.

4.1.9.5 Pyrazine-2-carboxylic acids

Tumova et al. (2005) found that the flavonolignan accumulation in callus culture of *S. marianum* after the elicitation with a solution of 5-brom-2-hydroxyphenyl-amide of 5-*tert*-butyl-6 chloropyrazine-2-carboxylic acid (24 h elicitation and concentration of 2.59×10^{-4} mol/l) on MS medium supplemented with NAA was increased by about 1039%. Also, elicitation with a solution of compound 3-methylamide 5-*tert*-butylpyrazine-2-carboxylic acid at a concentration of 3.71 × 10–7 mol·l-1 and within 72 h of elicitation on MS medium supplemented with 2,4-D caused an increase in flavonolignan production by 893% in cell suspension culture compared with control treatment. In another study, Tumova et al. (2010) found that the greatest increase of flavonolignan and taxifoline production was 0.080% in callus cultures of *S. marianum* with *N*-(3-iodo-4-methylphenyl) pyrazine-2-carboxamide (compound 1) at conc. 10 mg/100 ml after 6 h and 0.80% at conc. 1 mg/100 ml after 24 h. However, increased production of silychristin was recorded after 6 h elicitation with compound 1 at conc. 10 mg/100 ml (2 times higher compared to the control sample). In the case of elicitation with *N*-(3-iodo-4 methylphenyl)-5-*tert*-butyl pyrazine-2-carboxamide (compound 2), an increased production of silychristin was reached to 0.107% at the conc. 100 mg/100 ml after 72 h of elicitation (12 times compared to control), while silychristin content was 0.80% and 0.021% after 6 h at conc. 10 and 1 mg/100 ml (13 and 4 times control value, respectively). Also, Tumova et al. (2011) found that the highest content of silymarin complex (0.08%) in callus cultures of *S. marianum* was induced by 168 h elicitor application of a compound 5-(2 hydroxybenzoyl)-pyrazine-2-carboxamide at concentration 1.159×10^{-4} mol/l, compared with control (0.01%). In cell suspension cultures, the highest content of silydianin (the only detected compound) 0.11% was induced by 24 h elicitor application in concentration of 1.159×10^{-3} mol/l, compared with control (0.06%).

4.2 *Biotic Elicitation and Silymarin Production from In Vitro Culture of* Silybum marianum

Biotic elicitors have biological origin, often originated as a result of fungi, bacteria, virus, or herbivore infections (exogenous elicitors), and in some cases are released from the attacked plant by the action of enzymes of the pathogen (endogenous elicitors) (Angelova et al. 2006). The majority of biotic elicitors are recognized by specific receptors bound to the cell membrane. This stimulus is then transferred to the cell by a signal transduction system, producing changes that ultimately lead to the formation of phytoalexins (Baenas et al. 2014).

4.2.1 Yeast Extract and Chitosan

4.2.1.1 Yeast Extract

Yeast extract (biotic elicitor) has been used in plant tissue culture due to their ability to stimulate the defense mechanism, which leads to increase secondary metabolite production (Abraham et al. 2011). Results generated in the presence of different concentrations of yeast extract and chitosan on in vitro cultures of *S. marianum* are listed in Table 2.15.

In this respect, Sanchez-Sampedro et al. (2005b) reported that treatment of suspension cultures of *S. marianum* with a crude extract of yeast elicitor improved production of silymarin and caused the release of silymarin into the culture medium to a level about threefold higher than that of the control. However, treatment of cell culture with 100 and 200 μg/ml yeast extract for 48 h produced extensive browning and considerable loss of cell viability. Sanchez-Sampedro et al. (2007) reported that yeast extract contains several components that can elicit plant defense responses. Yeast extract acted on threonine and valine metabolism and promoted accumulation of choline and linolenic acid in 3-month-old undifferentiated hypocotyl callus of *S. marianum*.

Hasanloo et al. (2008b) found that the highest silymarin content (0.476 mg/g DW) was obtained by 2.5 mg/ml yeast extract, after 72 h elicitation in hairy roots of *S. marianum*. Also, Hasanloo et al. (2009) reported that treatment of *S. marianum* hairy roots with a crude extract of yeast elicitor improved the production of silymarin. Silymarin content reached a maximum of 0.47 mg/g DW by 72 h after culture (twofold higher than the control). They added that H_2O_2 content were increased 24 h treatment after YE elicitation. Rabie et al. (2010) showed that cells of *S. marianium* treated with yeast extract (40 mg/l) and 3 days incubation period resulted in the highest content of silymarin in callus cells to reach 8.6% over the

Table 2.15 Response of in vitro cultures of *S. marianum* to yeast extract and chitosan

Culture type	Biotic elicitor	Response	Reference
Cell suspensions (from 3-month-old hypocotyl callus)	Yeast extract (YE) (0.0, 5, 15, 50, and 100 μg/ml)	Silymarin in cells at level 50 μg/ml YE was threefold higher levels more than accumulated in the medium	Sanchez-Sampedro et al. (2005b)
	Yeast extract (50 μg/ml) Time course (0, 6, 12, 24, 48, 72, and 96 h)	At this level of YE, an increase in accumulation was noticed after 12 h of treatment reaching a maximum between 24 and 48 h and declining drastically	
Cell suspensions (from 3-month-old hypocotyl callus)	Yeast extract	Yeast extract acted on threonine and valine metabolism and promoted accumulation of choline and linolenic acid in cells Phenylpropanoid metabolism was altered by elicitation, but, depending on elicitor, different phenylpropanoid profile was produced	Sanchez-Sampedro et al. (2007)
Hairy roots	Yeast extracts (0, 0.5, 1, 2.5, and 5 mg/ml). Different exposure times (24, 48, and 72 h)	The highest growth index was obtained in media with 0.5 mg/l NAA. The highest silymarin content was obtained by 2.5 mg/ml yeast extract, after 72 h	Hasanloo et al. (2008b)
Hairy roots (30-day-old)	Yeast extracts (0, 0.5, 1, 2.5, and 5 mg/50 ml culture) Different exposure times (24, 48, 72, 96, and 120 h)	Silymarin content was increased after 72 h treatment with 2.5 mg/50 ml culture of yeast extract	Hasanloo et al. (2009)
Cell suspension culture from callus initiated from hypocotyl explants	Yeast extract (40, 45, 50, 55, and 60 mg/l) (3 and 28 day's incubation period)	Yeast extract at 40 mg/l gave the highest silymarin content (3-day incubation period)	Rabie et al. (2010)
Callus and cell suspension cultures	Yeast extracts (100 μg/ ml) Added to the 7-day-old cultures for 48 h	Yeast extract at conc. 100 μg ml^{-1} produced maximum silymarin accumulation after 48 h elicitation A twofold increase in silychristin content and threefold increase in silydianin content were observed when yeast extract was used as an elicitor	El-wekeel et al. (2012)

Cell cultures were developed from cotyledon and hypocotyl explants from 3-month-old callus Cultures were incubated at 25 °C and shaken at 90 rpm in darkness	Yeast extract (0.1% w/v)	SLM content of the yeast extract treated cell cultures increased significantly after 48 h No significant difference between the cell culture derived from cotyledon and hypocotyl in terms of silymarin content	Pourjabar et al. (2018)
Hairy roots (30 days old)	Medium molecular weight of chitosan at different concentrations (0, 10, 20, and 30 mg/50 ml) in different times (0, 12, 24, 48, 72, 96, and 120 h)	Total silymarin increased about 5.26-fold after 96 h of treatment with 30 mg/50 ml chitosan There was a gradual increase in dry weight content in treated hairy roots with 20 and 30 mg chitosan	Hasanloo et al. (2014)
Hairy root cultures	Treated by 30 mg/50 ml chitosan in different times (12, 24, 48, 72, 96, and 120 h)	The amount of silymarin increased after 96 h treatment	
Callus tissues derived from cotyledon (leaves) and hypocotyl	Chitosan (200, 400 and 800 mg/l)	Chitosan was most effective in accumulating the silydianin compared to untreated callus, SA, MeJA, and hairy root culture	Gabr et al. (2016)

control. El-wekeel et al. (2012) found that elicitation of cultured *S. marianum* cells with yeast extract (100 μg/ml) increased silychristin production from 0.11 (control) to 0.23 mg/g fresh weight and silydianin production from 0.01 to 0.03 mg/g fresh weight. Recently, Pourjabar et al. (2018) showed that yeast extract treatment (0.1% w/v) with callus tissues of *S. marianum* after 48 h elicitation increased the amount of silymarin production to 1.11-fold in comparison to the control. The highest amount of silymarin (approximately 1.38 mg/g dw) was obtained in cell suspension cultures derived from hypocotyl explant treated with yeast extract.

4.2.1.2 Chitosan

Chitosan is an exogenous biotic elicitor that is derived from the fungal cell wall (Montesano et al. 2003). Thus, chitosan may be involved in the signaling pathway for the biosynthesis of phenolic compounds and has been inducing chitinase and chitosanase, which are members of a group of plant pathogenesis-related (PR) proteins (Van Loon et al. 1994). These PR proteins can degrade the cell walls of some phytopathogens and, consequently, may play a role in host plant defense systems (Graham and Sticklen 1994). Moreover, chitosan may also induce plant immune systems (SAR), which is long lasting and often confers broad-based resistance to the different pathogens. SAR develops in uninfected parts of the plant; as a result, the entire plant is more resistant to a secondary infection such as antifungal chitinases, β-1, 3 glucanases, PR-1, and PR-5 (Sathiyabama and Balasubramanian 1998). Further studies have shown that chitosan induces the expression of various genes involved in plant defense responses such as a gene encoding PAL and protease inhibitors (Vander et al. 1998). The genetic studies support that chitosan may improve jasmonic acid (JA) pathways, since transcription activation of genes encoding PAL and protease inhibitors were induced by both JA and chitosan (Doares et al. 1995).

Elicitation of hairy root cultures of *S. marianum* with chitosan was studied (Hasanloo et al. 2014). The results indicated that chitosan had a positive effect on the biomass after 96 h. Total silymarin increased about 5.26-fold after 96 h of treatment with 30 mg/50 ml chitosan. Five different flavonolignans were isolated: taxifolin, silychristin, silydianin, silybin, and isosilybin, 0.133, 0.200, 0.120, 0.041, and 0.056 mg/g dry weight, respectively. The hairy roots which showed high silymarin content (grown with 30 mg/50 ml chitosan) were treated again with the same chitosan concentration and harvested at different time course (12, 24, 48, 72, 96, and 120 h). Analysis of these hairy roots showed significantly increasing amount of silymarin (0.705 mg/g dry weight) after 96 h treatment (2.9-fold higher than the control). Recently, Gabr et al. (2016) reported that chitosan treatments in callus cultures of *S. marianum* resulted in the lowest total phenolic accumulation, 0.72 μg/g DW at 800 mg/l concentration of chitosan compared to that of control which reached to 16.50 μg/g DW. Increasing concentrations of chitosan from 200 to 800 mg/l, the elicited callus showed gradual increase in total silymarin. The highest content of silymarin (5.77 μg/g DW) was recorded with 800 mg/l chitosan in comparison to the untreated callus (3.7 μg/g DW).

4.2.2 Fungal Elicitors

Plants often respond to microbial infection by producing antimicrobial compounds called phytoalexins, which by its turn triggers the cellular metabolic pathways toward the production of many intracellular hormones affecting the whole secondary metabolism process. Most studies done to date, reported that addition of fungal elicitors to cell cultures and pathogenic fungi has been used to enhance the synthesis of commercially important plant-derived compounds. Fungi, which mainly induce hypersensitive response in plant cells, result in activation of plant defense pathways and thereby increase phytoalexin production. Response of in vitro cultures of *S. marianum* to fungal elicitors is presented in Table 2.16.

El Sherif et al. (2013) found that among the supplemented fungal elicitors, to shoot cultures of *S. marianum, Penicilium chrysogenum* ATTC 2273 exerted the highest impact on total silymarin concentration to reach 3.885% dry weight compared to 1.842 and 0.602 for *Aspergillus niger* NRRL 3 and *Rhizopus stolonifer* NRRL 1472, respectively. The number of leaves, number of leaves/explant, and plant height decreased with increasing fungal applications. The type of fungi and their concentrations and exposure time have significant effect on the stimulation of silymarin production. In this respect, Hasanloo et al. (2013) showed that in the *S. marianum* hairy root cultures, the maximum silymarin production (0.34 mg/g

Table 2.16 Response of in vitro cultures of *S. marianum* to fungal elicitors

Culture type	Biotic elicitor (Fungal)	Response	Reference
In vitro shoot cultures	*Aspergillus niger* NRRL 3 (80 mg/l), *Rhizopus stolonifer* NRRL 1472 (80 mg/l), and *Penicillium chrysogenum* ATCC 2273 (20 mg/l)	*Penicilium chrysogenum* ATTC 2273 exerted the highest impact on total silymarin concentration	El-Sherif et al. (2013)
		The in vitro grown flowers showed 1.7 times more silymarin productivity as compared to that of the wild-grown plants	
Hairy roots	*Fusarium proliferatum, Aspergillus niger,* and *Rhizoctonia solani* (0, 10, and 20 mg/50 ml culture)	The maximum silymarin production for each of the fungi treatments were as follows: 10 mg *A. niger*/50 ml culture after 48 h, 20 mg/50 ml culture *F. proliferatum*, and 20 mg/50 ml culture *R. solani* after 72 h	Hasanloo et al. (2013)
Hairy root cultures (30 days old)	*Trichoderma* strains (KHB and G46–3) (0, 0.5, 1, 2, and 4 mg/ 50 ml culture	The highest silymarin accumulation was obtained in media elicited with 0.5 mg/50 ml cultures of *T. harzianum* strains (KHB and G46–3) after 120 h	Hasanloo et al. (2015)
	Strain KHB (0.5 mg/50 ml culture) Exposure times (0, 24, 48, 72, 96, and 120 h) for strain (KHB)	KHB strain, increased taxifolin, silychristin, isosilybin, and silydianin production after 120 h elicitation	

DW) was observed with 20 mg/50 ml culture *F. proliferatum* after 72 h elicitation. The flavonolignans of hairy roots treated with *F. proliferatum* included taxifoline (0.068 mg/g DW), silydianin (0.11 mg/g DW), silybin (0.021 mg/g DW), and isosilybin (0.015 mg/g DW) which were, respectively, 1.94-, 1.19-, 2.62-, and 1.25-fold greater than that the untreated cultures. A significant increase in dry weight of the cultures was observed in media supplemented with 10 mg/50 ml culture of *F. proliferatum*, and the highest biomass production (0.51 mg) was obtained 72 h after elicitation was significantly higher than the control (0.33 mg). Recently, Hasanloo et al. (2015) found that hairy root culture of *S. marianum* treated with 2 mg/50 ml culture of *T. harzianum* (KHB) resulted in the highest amount of dry weight (0.498 g/50 ml) that was 1.41-fold greater than the control. Silymarin production was decreased from 72 to 96 h and increased after 96 h and reached to 0.455 mg/g DW after 120 h. The H_2O_2 content in the control hairy root cultures remained lower than the treated cultures. There was significant enhancement in both peroxidase and ascorbate peroxidase activities in treated hairy roots reaching a peak after 72 h.

4.3 *Precursor Feeding and Silymarin Production from In Vitro Culture of* Silybum marianum

4.3.1 Coniferyl Alcohol, Tyrosine, and Phenylalanine

Precursors of biosynthetic pathways have been used in various plant cell cultures (callus, cell suspension, and hairy roots) to increase the yield of the desired product. Attempts to induce or increase the production of plant secondary metabolites, by supplying precursor or intermediate compounds, have been effective in different cultures of *S. marianum*. Factors such as the concentration and the time of addition of the precursor are to be considered when applying the precursor to the cell culture medium. Response of in vitro cultures of *S. marianum* to precursor feeding is shown in Table 2.17.

4.3.1.1 Coniferyl Alcohol

In the case of *S. marianum*, Tůmová et al. (2006) reported that the addition of coniferyl alcohol in cell suspension culture of *S. marianum* stimulated the highest taxifolin production after 48 h of treatment, about 554% compared to control. Later, Rady et al. (2014) found that coniferyl alcohol in concentration of 30 μM caused an increase in accumulation of silymarin in most callus cultures of *S. marianum*.

Table 2.17 Response of in vitro cultures of *S. marianum* to precursor feeding

Culture type	Precursor	Response	Reference
Cell suspension culture (from 3-month-old undifferentiated cotyledon calluses	Coniferyl alcohol (46 μM, 92 μM, and 184 μM) For 12, 24, 48, 72, and 168 hours of precursor application	A significant increase of silydianin was observed only after 72 h of the application of 46 μM coniferyl alcohol The highest taxifolin production was reached after 48 h of treatment (about 554% compared to control)	Tůmová et al. (2006)
		No other components of the silymarin complex (silychristin and silybin) were detected, neither in control samples nor after the precursor feeding	
Callus tissues	Coniferyl alcohol (10, 30, and 60 μM) for 3 months	Coniferyl alcohol in concentration of 30 μM caused an increase in accumulation of silymarin in most callus cultures	Rady et al. (2014)
Cell suspension culture from callus initiated from hypocotyl explants	L-Tyrosine (10, 20, 30, and 40 mg/l)	L-Tyrosine at 20 mg/l gave the highest silymarin content	Rabie et al. (2010)
	L- phenylalanine (10, 20, 30, and 40 mg/l)	Phenylalanine at all concentration gave silymarin content lower than control	
Hairy roots	L- phenylalanine (0, 1, 10, and 100 μM) (PHE) Hairy roots were harvested after 0, 24, 48, 72, 96, 120 h after addition of PHE	PHE 100 μM after 72 h was found optimal feeding condition for production of silymarin 1.84-fold higher than that of the control	Rahimi et al. (2011)
		Naringenin content indicated the similar pattern at 100 μM of PHE concentration 72 h after treatment which was 4.64-fold greater than untreated cultures. All three conc. of added PHE stimulated phenylalanine ammonia-lyase activity (PAL)	
Callus cultures derived from cotyledon	Phenylalanine (0, 1, 10, 25, and 100 μM)	Phenylalanine at 25 μM as precursor increased silymarin accumulation by 31%	Elhaak et al. (2016)

4.3.1.2 Tyrosine and Phenylalanine

Rabie et al. (2010) showed that the treatments which have tyrosine in concentration of 20 mg/l showed the highest production of silymarin in callus cells of *S. marianium* to reach 9.7% over the control, while phenylalanine treatments decreased significantly silymarin content than the control. In another study, feeding *S. marianum* hairy root cultures with phenylalanine (10 μM) increased the root biomass 1.89-fold compared to control 72 h after feeding. The presence of 100 μM phenylalanine in the media increased content of taxifoline, silydianin, and silybin to 4.16-, 2.44-, and 1.58-fold greater than that of untreated cultures after 72 h, respectively. However, the naringenin content was increased 4.64-fold greater than that of the untreated cultures at 100 μM concentration, 72 h after treatment. Moreover, all three concentrations of added PHE stimulated phenylalanine ammonia-lyase (PAL) activity and consequently supported the use of PHE as the substrate for the initial step of phenylpropanoid metabolism (Rahimi et al. 2011). Recently, Elhaak et al. (2016) reported that silymarin increased in callus cultures of *S. marianum* grown in medium with 25 μM of phenylalanine which enhanced accumulation of silymarin significantly by 31.4%, while accumulation of silymarin was decreased with increasing the concentration of phenylalanine up to 100 μM.

5 Conclusion

Milk thistle is one of the most important medicinal plants. Its fruits have been used medicinally for treatment of liver disorders. The present chapter includes an up-to-date published work on *S. marianum* using biotechnology approaches: micropropagation, direct, indirect regeneration, root initiation, and *Agrobacterium*-mediated transformation. Effects of abiotic, biotic, and precursor feeding on silymarin accumulation of several cell, callus, shoot, and hairy root cultures were extensively studied.

The current and recent review in this chapter presents different protocols which describe a simple, efficient, and productive culture technique for production of *S. marianum* plants using various explants cultured in vitro. Different explants derived from in vitro germinated seeds or in vivo grown silybum plants are reported for establishment of various cell, callus, and regeneration protocols in *S. marianum.* Selection and choice of the explant has a major role in determining the efficiency of in vitro culture and varies with the plant genotypes.

According to literature, it could be observed that the auxin NAA at levels ranging from 0.1 to 2.0 mg/l was the most auxins used for shoot culture initiation, multiplication, and rooting stages. BA at levels ranging from 0.25 to 5 mg/l was the most cytokinin used either alone or in combination with NAA in particular for initiation of shoot cultures. Also, it is obvious from previous works on initiation of root cultures that leaves, young shoots, nodal explants, and roots were used as explants for initiation of root cultures; NAA and/or IBA was recommended to induction of roots. It was reported that most studies have successfully established hairy root cultures via *Agrobacterium rhizogenes*-mediated transformation from different explants. Molecular analysis confirmed the presence and integration of the agrobacterium genes in the transgenic roots. However, transgenic hairy roots produced significantly higher amounts of flavonolignans which could be attributed to the overexpression of chsA gene at the root cells.

Elicitation strategies have been used in most reviewed reports to enhance the production of silymarin. It was repoted that different abiotic elicitors (medium composition, pH, growth regulators levels, photoperiod regime, and chemicals as well as stress factors) affect positively on the accumulation of silymarin in different silybum cultures. It was reported that the composition of the silymarin in in vitro cultures depends on the chemotype of silybum plants. It could be observed that the variations in genotypes caused changes in relative amounts of silymarin components from different origins and reflected in in vitro culture. Also, the specific Ca2+ chelator, EGTA, enhanced the silymarin content in cells by 200% and its secretion by 3–4 times. MeJA treatment (100 μM) resulted in increased amount of silymarin production to 1.23-fold in comparison to the control after 48 h. Silymarin accumulation was significantly increased 1.9-fold above the control in hairy roots of *S. marianum*.

Concerning the composition of the nutrient medium, glucose was found to be a better carbon source than fructose and sucrose in enhancing accumulation of silymarin in silybum cultures. Moreover, light treatment of root cultures induced more silymarin compared to dark treatment. However, with biotic elicitors, some reports revealed that a twofold increase in silychristin content and threefold increase in silydianin content in callus and suspension cultures were observed when yeast extract was used as a biotic elicitor. However, flowers of tissue culture raised plants treated with fungi showed 1.7 times more silymarin productivity as compared to that of the wild-grown plants. Hairy roots treated with *F. proliferatum* contained silymarin more twofold increase than that the untreated cultures. Exogenous supply of the biosynthetic precursor phenylalanine to the culture medium increases the yield of silymarin accumulation by 31% in callus cultures. Silymarin content in hairy roots was 1.48-fold higher than that of the control when phenylalanine was added to the medium. Therefore, selection of cell lines with suitable biochemical and physiological characteristics is an important approach to improve the productivity for silymarin.

6 Future Aspects

Several researchers have been successfully reported establishment of protocols for cell, tissue, and hairy root cultures of *S. marianum*. Achieving higher frequencies of somatic embryos proliferation in *S. marianum* tissue culture for allowing long-term culture maintenance and high, productive frequencies of genetic transformation. Also, genetic transformation with genes involved in the biosynthesis of silymarin in in vitro cultures is still important research areas for production of silymarin. Efforts among various disciplines will be needed to accelerate the development of an efficient system for industrial-scale metabolite production in the near future. However, until now, the yields obtained from cell cultures are still low, and new strategies as the use of bioreactors for increasing silybum cell biomass and silymarin are still on progress, and, therefore, much work will have to be done to provide high amounts of this important bioactive compound. The understanding of biosynthetic pathways of flagogonins in in vitro cultures of *S. marianum* is still preliminary, and, therefore, the knowledge of the regulation of this pathway concerned with the level of products, enzymes and genes, as well as transport and compartmentation is still required. It will also be interesting for future studies to test activities of key enzymes of the flavonolignan pathway in cell cultures of *S. marianum*. Recently, the emergence of synthetic biology offers the possibility of improving the production of useful products through the design and engineering of complex biological systems.

References

Abbasi BH, Khan MA, Mahmood T, Ahmad M, Chaudhary MF, Khan Mir A (2010) Shoot regeneration and free-radical scavenging activity in *Silybum marianum* L. Plant Cell Tissue Organ Cult 101:371–376

Abbasi BH, Ali H, Yucesan B, Saeed S, Rehman K, Khan MA (2016) Evaluation of biochemical markers during somatic embryogenesis in *Silybum marianum* L. 3 Biotech 6:71

Abraham F, Bhatt A, Keng CL, Indrayanto G, Sulaiman SF (2011) Effect of yeast extract and chitosan on shoot proliferation, morphology and antioxidant activity of *Curcuma mangga* in *vitro* plantlets. Afr J Biotechnol 10(40):7787–7795

Ahmad N, Fatima N, Ahmad I, Anis M (2015) Effect of PGRs in adventitious root culture *in vitro*: present scenario and future prospects. Rend Fis Acc Lincei 26:307–321

Al-Hawamdeh FM, Shibli RA, Al-Qudah TS (2013) *In-vitro* production of silymarin from *Silybum marianum* L. Med Aromat Plants S1:001

Al-Hawamdeh FM, Shibli RA, Al-Qudah TS (2014) *In vitro* propagation of *Silybum marianum* L. Jordan J Agric Sci 10(1):120–129

Angelova Z, Georgiev S, Roos W (2006) Elicitation of plants. Biotechnol Biotechnol Equip 20:72–83

Baenas N, Garcia-Viguera C, Moreno DA (2014) Elicitation: a tool for enriching the bioactive composition of foods. Molecules 19(9):13541–13563

Bais HP, Loyola-Vargas VM, Flores HE, Vivanco JM (2001) Root-specific metabolism: the biology and biochemistry of underground organs. In Vitro Cel Dev Biol Plant 37:730–741

Bekheet SA (2015) Effect of drought stress induced by mannitol and polyethylene glycol on growth and silymarin content of milk thistle callus cultures. World J Pharm Res 4(8):116–127

Bekheet SA, Taha HS, Gabr AMM (2013) Protocol for *in vitro* morphogenesis and hairy root cultures of Milk thistle (*Silybum marianum* L. Gaertn). J Appl Sci Res 9(1):860–866

Bekheet SA, El-Bahr MK, Ali SA, Hamed MA (2014) Callus production of globe artichoke and milk thistle: *in vitro* hypolipidemic and antioxidant activities. World J Pharm Res 3(4):01–17

Belchi-Navarro S, Pedreno MA, Corchete P (2011) Methyl jasmonate increases silymarin production in *Silybum marianum* (L.) Gaernt cell cultures treated with β-cyclodextrins. Biotechnol Lett 33:179–184

Biedermann D, Vavrikov E, Cvak L, Kren V (2014) Chemistry of silybin. Nat Prod Rep 31:1138–1157

Boerjan W, Ralph J, Baucher M (2003) Lignin biosynthesis. Annu Rev Plant Biol 54:519–546

Bota C, Deliu C (2011) The effect of copper sulphate on the production of flavonoids in *Digitalis lanata* cell cultures. Farmacia 59(1):11–118

Cacho M, Moran M, Corchete P, Fernandez-Tarrago J (1999) Influence of medium composition on the accumulation of flavonolignans in cultured cells of *Silybum marianum* L. Gaertn. Plant Sci 144:77–84

Chen Z, Silva H, Klessig DF (1993) Active oxygen species in the induction of plant systemic resistance by salisylic acid. Science 262:1883–1886

Cimino C, Cavalli SV, Spina F, Natalucci C, Priolo N (2006) Callus cultures for biomass production of milk thistle as a potential source of milk clotting peptidases. Electron J Biotechnol 9(3):1–4

Corchete P (2008) *Silybum marianum* (L.) Gaertn: the source of silymarin. In: Ramawat KG, Merillon JM (eds) Bioactive molecules and medicinal plants chapter 6, pp 123–148

Cui XH, Chakrabarty D, Lee EJ, Paek KY (2010) Production of adventitious roots and secondary metabolites by *Hypericum perforatum* L. in a bioreactor. Bioresour Technol 101:4708–4716

Dewick PM (2002) Medicinal natural products. A Biosynthetic Approach. Second Edition Chichester, UK: Wiley

Doares SH, Syrovets T, Weiler EW, Ryan CA (1995) Oligogalacturonides and chitosan activate plant defensive genes through the octadecanoid pathway. Proc Natl Acad Sci U S A 92:4095–4098

Dubreuil-Maurizi C, Vitecek J, Marty L, Branciard L, Frettinger P, Wendehenne D, Meyer AJ, Mauch F, Poinssot B (2011) Glutathione deficiency of the *Arabidopsis* mutant pad2-1 affects oxidative stress-related events, defense gene expression, and the hypersensitive response. Plant Physiol 157(4):2000–2012

Eari S, Aghdasi M, Ahmadzadeh E, Mianabadi M (2017) Influence of plant growth regulators on callus induction, silymarin production and antioxidant activity in Milk Thistle (*Silybum marianum* L. Gaertn.) under tissue culture medium. J Med Plants By-Products 1:59–69

Elhaak M, Zayed Mattar M, Gad D, Dietz K (2016) Optimization of *Silybum marianum* L. callus production and magnifying callus silymarin accumulation by elicitors or precursors. Intr J Adv Pharm Biol Chem 5(2):148–163

Elsharnouby ME, Hassan FAS (2018) Improvement of silymarin content in cell cultures of *Silybum marianum* by copper sulphate elicitor. Acta Sci Pol Hortorum Cultus 17(2):105–114

El-Sherif F, Khattab S, Ghoname E, Salem N, Radwan K (2011) Effect of gamma irradiation on enhancement of some economic traits and molecular changes in *Hibiscus Sabdariffa* L. Life Science J 8(3):220–229

El-Sherif F, Khattab S, Ibrahim AK, Ahmed SA (2013) Improved silymarin content in elicited multiple shoot cultures of *Silybum marianum* L. Physiol Mol Biol Plants 19(1):127–136

El-wekeel A, AbouZid S, Sokkar N, Elfishway A (2012) Studies on flavanolignans from cultured cells of *Silybum marianum*. Acta Physiol Plant 34:1445–1449

Feveriro P, Carbal JMS, Fonceca MMR, Novais JM, Salom M (1986) Callus and suspension culture of *Silybum marianum* biosynthesis of proteins with clotting activity. Biotechnol Lett 8(1):19–24

Gabay R, Plitmann U, Danin A (1994) Factors affecting the dominance of *Silybum marianum* L. (Asteraceae) in its specific habitats. Flora 189:201–206

Gabr AMM, Ghareeb H, El Shabrawi HM, Smetanska I, Bekheet SA (2016) Enhancement of silymarin and phenolic compound accumulation in tissue culture of Milk thistle using elicitor feeding and hairy root cultures. J Genet Eng Biotechnol 14(2):327–333

Gamborg OL, Miller RA, Ojima K (1968) Nutrient requirements of suspension culture of soybean root cells. Exp Cell Res 50 (1):151–158

Graham LS, Sticklen MB (1994) Plant chitinases. Can J Bot 72:1057–1083

Han YS, Van der Heijden R, Verpoorte R (2001) Biosynthesis of anthraquinones in cell cultures of the Rubiaceae. Plant Cell Tissue Organ Cult 67:201–220

Hasanloo T, Rahnama H, Sepehrifar R, Shams MR (2008a) The influence of yeast extract on the production of flavonolignans in hairy root cultures of *Silybum marianum* L. Gaertn. IFMBE Proc 21:358–361

Hasanloo T, Khavari-Nejad RA, Majidi E, Shams MR (2008b) Flavonolignan production in cell suspension culture of *Silybum marianum*. Pharm Biol 46(12):876–882

Hasanloo T, Sepehrifar R, Rahnama H, Shams MR (2009) Evaluation of the yeast-extract signaling pathway leading to silymarin biosynthesis in milk thistle hairy root culture. World J Microbiol Biotechnol 25:1901–1909

Hasanloo T, Ahmadi M, Nekoei SMK, Jouzani GS (2013) Improvement of silymarin production in hairy root cultures of *Silybum marianum* (L.) Gaertn using fungal elicitors. Romanian Biotechnol Lett 18(3):8221–8232

Hasanloo T, Eskandari S, Najafi F (2014) Chitosan (middle-viscous) as an effective elicitor for silymarin production in *Silybum marianum* hairy root cultures. Res J Pharm 1:9–13

Hasanloo T, Eskandari S, Kowsari M (2015) *Trichoderma* strains- *Silybum marianum* hairy root cultures interactions. Res J Pharm 2(2):33–46

Hidalgo D, Martınez-Marquez A, Cusido R, Bru-Martınez R, Palazon J, Corchete P (2017) *Silybum marianum* cell cultures stably transformed with *Vitis vinifera* stilbene synthase accumulate *t*-resveratrol in the extracellular medium after elicitation with methyl jasmonate or methylated *β*-cyclodextrins. Eng Life Sci 17:686–694

Iqbal SM, Srivastava PS (2000) *In vitro* micropropagation of *Silybum marianum* L. from various explants and silybin Content. J Plant Biochem Biotechnol 9:81–87

Iri S, Aghdasi M, Mianabadi M (2013) Induction of root formation to produce silymarin in *Silybium marianum* plant in tissue culture condition. J Plant Process Funct 2(2):1–12

Jabeen Z, Rahman KU, Zia MA, Jhan N (2019) Establishment of an efficient and reproducible *in vitro* protocol for callogenesis of *Silybum marianum*. Int J Biosci 14(1):402–410

Jamwal K, Bhattacharya S, Puri S (2018) Plant growth regulator mediated consequences of secondary metabolites in medicinal plants. J Appl Res Med Aromatic Plants 9:26–38

John SA, Koperuncholan M (2012) Direct root regeneration and indirect organogenesis in *Silybum marianum* and preliminary phytochemical, antibacterial studies of its callus. Int J Pharm 2(2):392–400

Khalili M, Hasanloo T, Tabar SKK, Rahnama H (2009) Influence of exogenous salicylic acid on flavonolignans and lipoxygenase activity in the hairy root cultures of *Silybum marianum*. Cell Biol Int 33:988–994

Khalili M, Hasanloo T, Kazemi Tabar SK, Sepehrifar R (2010a) Effect of salicylic acid on antioxidant activity in Milk Thistle hairy root cultures. J of Medicinal Plants 9(35):51–60

Khalili M, Hasanloo T, SKk T (2010b) Ag^+ enhanced silymarin production in hairy root cultures of *Silybum marianum* (L.) Gaertn. Plant Omics J 3(4):109–114

Khan MA, Abbasi BH, Ahmed N, Ali H (2013) Effects of light regimes on *in vitro* seed germination and silymarin content in *Silybum marianum*. Ind Crop Prod 46:105–110

Khan MA, Abbasi BH, Khan Z, Shi ZK (2014) Thidiazuron enhanced regeneration and silymarin content in Silybum marianum L. Pak J Bot 46(1):185–190

Khan MA, Abbasi BH, Shah NA, Yucesan B, Ali H (2015a) Analysis of metabolic variations throughout growth and development of adventitious roots in *Silybum marianum* L. (Milk thistle), a medicinal plant. Plant Cell Tissue Organ Cult 123:501–510

Khan MA, Abbasi BH, Ali H, Ali M, Adil M, Hussain I (2015b) Temporal variations in metabolite profiles at different growth phases during somatic embryogenesis of *Silybum marianum* L. Plant Cell Tissue Organ Cult 120:127–139

Kim Y, Wyslouzil BE, Weathers PJ (2002) Secondary metabolism of hairy root cultures in bioreactors. In Vitro Cell Dev Bio Plant 38:1–10

Kim NC, Graf TN, Sparacino CM, Wani MC, Wall ME (2003) Complete isolation and characterization of silybins and isosilybins from milk thistle (Silybum marianum). Org Biomol Chem 1:1684–1689

Kim JH, Baek MH, Chung BY, Wi SG, Kim JS (2004) Alterations in the photosynthetic pigments and antioxidant machineries of red pepper (Capsicum annuum L.) seedlings from gamma-irradiated seeds. J Plant Biol 47:314–321

Lee DYW, Liu YZ (2003) Molecular structure and stereochemistry of silybin A,silybin B, isosilybin A, and isosilybin B, isolated from *Silybum marianum* (milkthistle). J Nat Prod 66:1171–1174

Liu S-Q, Cai Q-G (1990) Callus formation from protoplasts and plant regeneration from tissue culture of *Silybum marianum* Gaertn. Acta Bot Sin 32(1):19–25

Lv Y, Gao S, Xu S, Du G, Zhou J, Chen J (2017a) Spatial organization of silybin biosynthesis in milk thistle [Silybum marianum (L.) Gaertn]. The Plant J 92(6):995–1004

Lv Y-W, Rui-Jie W, Lv Y-W, Ze-Shang Y, Wang Y-J (2017b) *In vitro* propagation of *silybum marianum* (l.) gaertn. and genetic fidelity assessment of micropropagated plants. Pak J Bot 49(2):673–680

Madrid E, Corchete P (2010) Silymarin secretion and its elicitation by methyl jasmonate in cell cultures of *Silybum marianum* is mediated by phospholipase D-phosphatidic acid. J Exp Bot 61(3):747–754

Manaf HH, Rabie KAE, Abd El-Aal MS (2009) *In vitro* callus formation and plant regeneration of *S. marianum* L. Gaertin. Annals Agric Sci Ain Shams Univ 54(2):283–289

Manaf HH, Rabie KAE, Ibrahim IS, Eltantawy MEI (2016) Effects of BAP, NAA and NaCl on growth characters and silymarin concentration of Silybum marianum callus. J Biol Chem Environ Sci 11(4):139–151

Mittler R, Vanderauwera S, Gollery M, Breusegem FV (2004) Abiotic stress series. Reactive oxygen gene network of plants. Trends Plant Sci 9(10):490–498

Montesano M, Brader G, Palva ET (2003) Pathogen derived elicitors: searching for receptors in plants. Mol Plant Pathol 4:73–79

Morazzoni P, Bombardelli E (1995) *Silybum marianum* (Carduus marianum). Fitoterapia 66:3–42

Murashige T, Skoog F (1962) A revised medium for rapid growth and bioassays with tobacco tissue cultures. Physiol Plant 15:473–497

Nikolova MT, Ivancheva SV (2005) Quantitative flavonoid variations of *Artemisia vulgaris* L. and *Veronica chamaedrys* L. in relation to altitude and polluted environment. Acta Biol Szeged 49:29–32

Owolabi LO, Yupanqui CT, Siripongvutikorn S (2018) Enhancing secondary metabolites (emphasis on phenolics and antioxidants) in plants through elicitation and metabolomics. Pak J Nutr 17:411–420

Petrussa E, Braidot E, Zancani M, Peresson C, Bertolini A, Patui S, Vianello A (2013) Plant flavonoids—biosynthesis, transport and involvement in stress responses. Int J Mol Sci 14:14950–14973

Poppe L, Petersen M (2016) Variation in the flavonolignan composition of fruits from different *Silybum marianum* chemotypes and suspension cultures derived therefrom. Phytochemistry 131:68–75

Pourjabar A, Mohammadi SA, Ghahramanzadeh R, Gh S (2012) Effect of genotype, explant type and growth regulators on the accumulation of flavonoides of (*Silybum marianum L.*) in *in vitro* culture. Int J Biol Biomol Agric Food Biotech Eng 6(7):514–516

Pourjabar A, Azimi MR, Mostafaie A, Kahrizi D, Cheghamirza K (2018) Proteome analysis of Milk thistle (Silybum marianum L.) cell suspension cultures in response to methyl jasmonate and yeast extract elicitors. Appl Ecol Environ Res 17(1):547–560

Rabie KAE, Abd El-Aal MS, Manaf HH (2010) Enhanced silymarin accumulation as influence of medium composition in cell suspension cultures of *Silybum marianum* (L.) Gaertin. J of Plant Production 1(2):319–332

Rady MR, Matter MA, Ghareeb HA, Hanafy MS, Saker MM, Eid SA, Hammoda FM, Imbaby SI, Nazief NH (2014) *In vitro* cultures of *Silybum marianum* and silymarin accumulation. J Genetic Eng Biotechnol 12:75–79

Rahimi S, Hasanloo T (2016) The effect of temperature and pH on biomass and bioactive compounds production in *Silybum marianum* hairy root cultures. Res J Pharm 3(2):53–59

Rahimi S, Hasanloo T, Bihamta MR (2010) Enhanced production of silymarin by Ag+ elicitor in cell suspension cultures of *Silybum marianum*. Pharm Biol 48(6):708–715

Rahimi S, Hasanloo T, Najafi F, Khavari-Nejad RA (2011) Enhancement of silymarin accumulation using precursor feeding in *Silybum marianum* hairy root cultures. Plant Omic J 4(1):34–39

Rahnama H, Hasanloo T, Shams MR, Sepehrifar R (2008) Silymarin production by hairy root culture of *Silybum marianum* (L.) Gaertn. Iran J Biotechnol 6(2):113–118

Rahnama H, Razi Z, Dadgar MN, Hasanloo T (2013) Enhanced production of flavonolignans in hairy root cultures of *Silybum marianum* by over-expression of chalcone synthase gene. J Plant Biochem Biotechnol 22(1):138–143

Riasat R, Riasat Z, Abbasi BH, Liu C, Khan MA (2015) *Silybum marianum*: adventitious roots induction along with free radical scavenging activity. J Plant Bio Res 4(1):12–21

Russo GL, Russo M, Spagnuolo C, Tedesco I, Bilotto S, Iannitti R, Palumbo R (2014) Quercetin a pleiotropic kinase inhibitor against cancer. Cancer Treat Res 159:185–205

Sagar SM (2007) Future directions for research on *Silybum marianum* for cancer patients. Integr Cancer Ther 6:166–173

Sanchez-Sampedro MA, Fernandez-Tarrago J, Corchete P (2005a) Enhanced silymarin accumulation is related to calcium deprivation in cell suspension cultures of *Silybum marianum* (L.) Gaertn. J Plant Physiol 162:1177–1182

Sanchez-Sampedro MA, Fernandez-Tarrago J, Corchete P (2005b) Yeast extract and methyl jasmonate-induced silymarin production in cell cultures of *Silybum marianum* (L.) Gaertn. J Biotechnol 119(1):60–69

Sanchez-Sampedro A, Kim HK, Choi YH, Verpoorte R, Corchete P (2007) Metabolomic alterations in elicitor treated *Silybum marianum* suspension cultures monitored by nuclear magnetic resonance spectroscopy. J Biotechnol 130:133–142

Sathiyabama M, Balasubramanian R (1998) Chitosan induces resistance component in *Arachis hipogaea* leaf rust caused by *Puccinia arachidis*. Speg Crop Prot 17:307–313

Schenk RU, Hildebrandt AC (1972) Medium and techniques for induction and growth of monocotyledonous and dicotyledonous plant cell cultures. Can J Bot 50:199–204

Seibert M, Kadkade PG (1980) Environmental factors. A.Light. In: Staba EJ (ed) Plant tissue culture as a source of Biochemicals. CRC Press, Boca Raton, pp 123–141

Shah J, Klessig DF (1999) Salicylic acid: signal perception and transduction. In: Libbenga K, Hall M, Hooykaas PJJ (eds) Biochemistry and molecular biology of plant hormones. Elsevier, Oxford, pp 513–541

Shokrpour M, Mohammadi SA, Moghaddam M, Ziai SA, Javanshir A (2007) Variation in flavonolignan concentration of milk thistle (Silybum marianum) fruits grown in Iran. J Herbs Spices Med Plants 13:55–69

Shualev V, Leon J, Raskin I (1995) Is salicylic acid a translocated signal of systemic acquired resistance in tobacco. Plant Cell 7:1691–1701

Sivakumar G, Alagumanian S, Rao MV (2006) High frequency *in vitro* multiplication of *Centella asiatica*: an Important Industrial Medicinal Herb. Eng Life Sci 6:597–601

Sivanandhan G, Dev GK, Jeyaraj M, Rajesh M, Arjunan A, Muthuselvam M, Manickavasagam M, Selvaraj N, Ganapathi A (2013) Increased production of withanolide A, withanone, and withaferin A in hairy root cultures of *Withania somnifera* (L.) Dunal elicited with methyl jasmonate and salicylic acid. Plant Cell Tissue Organ Cult 114:121–129

Tumova L, Gallova K, Rimakova J (2001) Silybum marianum. In vitro. Ceska Slov Farm 53:135–140

Tumova L, Gallova K, Rimakova J, Dolezal M, Tuma J (2005) The effect of substituted amides of pyrazine-2-carboxylic acids on flavonolignan production in *Silybum marianum* culture in vitro. Acta Physiol Plant 27(38):357–362

Tůmová L, Řimáková J, Tůma J, Dušek J (2006) *Silybum marianum* in vitro-flavonolignan production. Plant Soil Environ 52(10):454–458

Tůmová L, Tůma J, Megušar K, Doležal M (2010) Substituted pyrazinecarboxamides as abiotic elicitors of flavolignan production in *Silybum marianum* (L.) Gaertn cultures *in vitro*. Molecules 15:331–340

Tumova L, Tuma J, Dolezal M (2011) Pyrazinecarboxamides as potential elicitors of flavonolignan and flavonoid production in *Silybum marianum* and *Ononis arvensis* cultures *in vitro*. Molecules 16:9142–9152

Van Loon LC, Pierpoint WS, Voller T, Conejero V (1994) Recommendations for naming plant pathogenesis related protein. Plant Mol Biol Report 12:245–264

Vander P, Varum KM, Domard A, Gueddari NEE, Moerschbacher BM (1998) Comparison of the ability of partially *N*-acetylated chitosans and chitooligosaccharides to elicit resistance reactions in wheat leaves. Plant Physiol 118:1353–1359

Vanisree M, Lee CY, Lo SF, Nalawade SM, Lin CY, Tsay HS (2004) Studies on the production of some important secondary metabolites from medicinal plants by tissue culture. Bot Bull Acad Sin 45:1–22

Vasconsuelo A, Boland R (2007) Molecular aspects of the early stages of elicitation of secondary metabolites in plants. Plant Sci 172:861–875

Vildová A, Hendrychová H, Kubeš J, Tůmová L (2014) Influence of $AgNO_3$ treatment on the flavonolignan production in cell suspension culture of *Silybum marianum* (L.) Gaertn. Int J Biol Biomol Agric Food Biotech Eng 8(8):959–962

Wi SG, Chung BY, Kim JH, Baek MH, Yang DH, Lee JW, Kim JS (2005) Ultrastructural changes of cell organelles in Arabidopsis stem after gamma irradiation. J Plant Biol 48(2):195–200

Yoshikawa T, Furuya T (1987) Saponin production by cultures of *Panax ginseng* transformed with *Agrobacterium rhizogenes*. Plant Cell Rep 6:449–453

Younas M, Drouet S, Nadeem M, Giglioli-Guivarch N, Hano C, Bilal Haider Abbasi BH (2018) Differential accumulation of silymarin induced by exposure of *Silybum marianum* L callus cultures to several spectres of monochromatic lights. J Photochem Photobiol B: Biol 184:61–70

Zhang SL, Zhang TZ, Yang SH (2014) Establishment of culture system of Silybum marianum hairy roots and determination of silybin. China J Chinese Mater Med Zhongguo Zhong Yao Za Zhi 39(11):2005–2010

Zhao J, Davis LC, Verpoorte R (2005) elicitor signal transduction leading to production of plant secondary metabolites. Biotechnol Adv 23:283–333

Zhao J, Hu Q, Zhu WH (2001) Enhanced catharanthine production in Catharanthus roseus cell cultures by combined elicitor treatment in shake flasks and bioreactors. Enzyme Microb Technol 28:673–681

Chapter 3
Plant Biotechnology and Foxglove

Contents

M. R. Rady, *Plant Biotechnology and Medicinal Plants*,
https://doi.org/10.1007/978-3-030-22929-0_3

Abstract The genus *Digitalis*, commonly known as the "foxglove," is one of the most important medicinal plants belonging to the family of the Plantaginaceae. *Digitalis* ssp. is used exclusively for the isolation of individual cardiac glycosides, principally digoxin and lanatoside C. Cardenolides from *Digitalis* are well known and frequently used as pharmaceuticals for the treatment of heart disorders. An overview of recent studies which have been used using various approaches of plant tissue, organ culture, regeneration, cryopreservation, and transformation of *Digitalis* ssp. is presented. This chapter summarizes the recent research work of various in vitro culture of *Digitalis* ssp. through abiotic, biotic elicitors, and precursor feeding applied to *Digitalis* cultural system and their stimulating effects on the accumulation of cardenolides.

Keyword Digitalis ssp. · Biosynthesis · Cell and callus culture · Regeneration · Cryopreservation · Transformation · Hairy root cultures · Elicitation · Cardenolides

1 Introduction

Species in the genus *Digitalis* L. are the main sources of cardenolides. Cardiac glycosides are metabolites of major interest in pharmaceutical industries. All *Digitalis* species are biennial or perennial herbs, rarely small shrubs with simple, alternate leaves, which are often crowded in basal rosettes (Fig. 3.1). Flowers are zygomorphic and arranged in terminal, bracteate racemes and vary in color with species, from purple to pink, white, and yellow (Brauchler et al. 2004). Digitalis species, specifically *D. lanata* and *D. purpurea*, have been intensively studied for their medicinal and economical values because they contain cardiac glycosides, which regulate heart rhythm and can be used to treat myocardial infarction, angina, cardiac dysfunction, hypertrophy, and arterial hypertension and increase the force of systolic contractions (Roden 2001; Yücesan et al. 2018). Moreover, the use of *Digitalis* leaves for effective cancer chemotherapy, in particular, for treating prostate and breast cancer (Gurel et al. 2011; Elbaz et al. 2012; Yücesan et al. 2014). Cardiovascular and cancer diseases are the first causes of death in the world. *D. purpurea* is a medicinal plant that produces secondary metabolites, like digoxin and digitoxin, which are employed in therapies against heart failure (Pérez-Alonso et al. 2018).

Since many *Digitalis* species are biennial, seed setting takes 1 additional year after basal leaf formation. This plant faces some problems in natural populations like seasonal variations and mineral soil characteristics, rhizosphere communities, misidentification, genetic and phenotypic variability, poor germination of seeds, and extract variability which can inherent in its extracts also affect cardenolide production in the plant. Therefore, there has been considerable interest in creating an economically feasible biotechnological process as an alternative method for improving this plant. Cardiotonic glycosides are extracted mostly from leaves of Digitalis plants. Commercial production of cardenolides by traditional agriculture is an inefficient process and can be affected by climatic and soil conditions. Strategies, based on in vitro culture methods, have been extensively studied to

Fig. 3.1 Digitalis plant

improve the production of specific plant-derived chemicals (Pérez-Alonso et al. 2009). In this chapter, recent studies using various biotechnological approaches such as callus, suspension, plant regeneration (viz., somatic embryogenesis and organogenesis), cryopreservation, transformation, and hairy root were used and have been reviewed thoroughly. However, the chapter focuses specifically on the effect of different abiotic, biotic, and precursor feeding on the production cardenolides from different cultures of *Digitalis* ssp.

2 Biosynthesis of Cardenolides in *Digitalis* ssp.

The most prominent compounds formed in the genus *Digitalis* are the cardenolides. Cardenolides are a group of plant secondary metabolites, many of which possess potent pharmacological and other bioactive properties. Perez-Bermudez et al. (2010) reported that cardiac glycosides are widely used as inotropic drugs to improve hemodynamics in patients with compromised cardiac function.

Formation of progesterone
Cardenolide biosynthesis requires the transformation of sterols to pregnenolone. Cholesterol is the direct sterol precursor of cardenolides, converted to pregnenolone by a P450scce (cholesterol side-chain cleavage enzyme). Then, this pregnene is converted by Δ5-3β-hydroxysteroid dehydrogenase into isoprogesterone, which is isomerized to progesterone. Progesterone 5β-reductase catalyzes the 5β-reduction of progesterone to 5β-pregnan-3,20-dione and is considered to be the first committed step in the branch pathway leading to cardenolides.

Formation of digitoxigenin

In a second stereospecific reaction, 3β-hydroxysteroid-5β-oxidoreductase converts 5β-pregnan-3,20-dione into 5β-pregnan-3β-ol-20-one. Sequential hydroxylations at C14 and C21 and lactone ring formation at C17 lead to aglycone digitoxigenin (Perez-Bermudez et al. 2010). P5βR is considered to be a key enzyme in cardenolide biosynthesis as: it is the first stereospecific enzyme of the pathway leading to 5β-configured derivatives; it appears to be the initial step in cardenolide biosynthesis; and it is the only known enzyme which is exclusively present in shoot cultures and young *Digitalis* plants (Gartner and Seitz 1993). Kreis et al. (1998) reported that about 20 enzymes which probably affect the formation of cardenolides have been identified in recent years. Among these are enzymes responsible for cardenolide formation and storage, as well as enzymes responsible for removing precursors from the cardenolide pathway. Only a few of them have been purified and just one, namely, progesterone 5 J3-reductase, has been partially sequenced. The discovery of enzymes which have not yet been described may help to clarify the cardenolide pathway(s).

Other steps of the biosynthesis, including the formation of the sugar side chain, have not provided any new insights over the past 20 years, although pregnane 2β-hydroxylation was described (Pádua et al. 2012). In addition to the 12β- and 16β-hydroxylations already known (Luckner and Wichtl 2000), no attempts to isolate the respective enzymes and/or genes were reported. Biosynthetic routes leading to the cardenolides are outlined in Fig. 3.2.

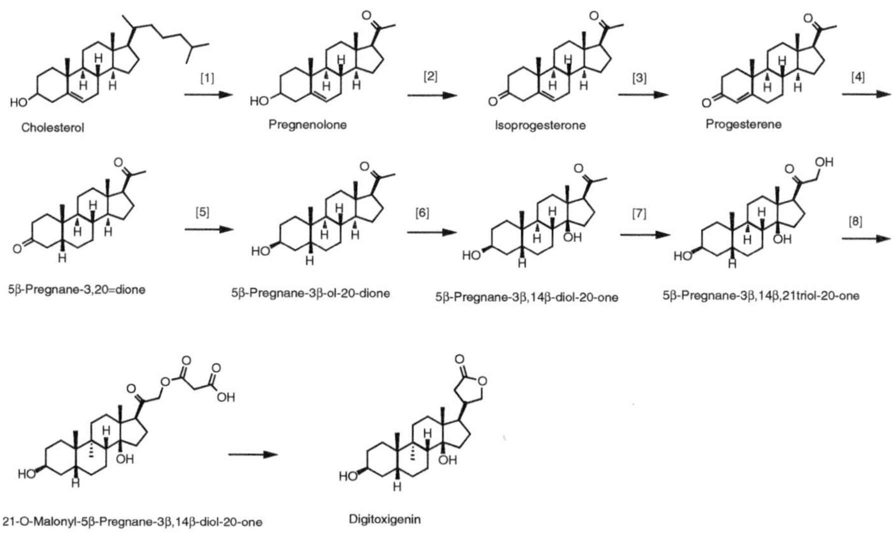

Fig. 3.2 Postulated biosynthetic pathway of cardenolide formation in *Digitalis*. 1 Sterol side-chain cleavage enzyme (SCCE), 2 NAD:3β-hydroxysteroid dehydrogenase (3β-HSD). 3 Δ4,5–3-ketosteroid-isomerase (3-KSI), 4 progesterone-5β-reductase (P5βR), 5 NADH:3β-hydroxysteroid dehydrogenase (3β-HSD), 6 steroid 14β-hydroxylase (unknown), 7 steroid 21β-hydroxylase (unknown), and 8 malonyl coenzyme A:21-hydroxypregnane 21-O-malonyltransferase (21MaT). (Adapted from Kreis 2017)

3 In Vitro Culture of *Digitalis* ssp.

3.1 Cell and Callus Cultures of Digitalis *ssp.*

The establishment of callus and cell cultures from *Digitalis* ssp. facilitates investigations on metabolism and development independently from the complex organism. Starting from callus it was possible to generate permanent suspension cultures submerged in liquid media. Callus formation was induced on explants from different parts of seedlings as well as of adult plants of different ages of several *Digitalis* species. The callus cultures differed in growth rate, morphology, color, and the tendency of organ differentiation, according to medium, growth regulators and kind of explants used for cultivation. Studies on cell and callus cultures induction of *Digitalis* spp. are presented in Table 3.1.

Production of haploid callus from the anthers of *D. purpurea* flowering shoots was obtained after a culture period of 8 weeks using different culture media (Corduan and Spix 1975). Anther-derived callus was produced only when the flowering shoots were kept at 4 °C for 2–12 days before the anthers were isolated. The best media for callus formation were white medium with 10 mg/l 2,4-D or Nitsch medium with 5.0 mg/l 2,4-D, producing 16.6% and 11.0% callus formation in continuous light or dark respectively. However, it was found no significant differences in callus production from anthers between populations from different regions. Perez-Bermudez et al. (1984) reported that MS medium with NAA (1.0 mg/l) + BA (2.0 mg/l) was the best for callus formation from leaf segment of *D. obscura* seedlings. In that medium, callugenic explant % was 23. BA showed a clear synergism with auxins not only in the callus induction but also in their growth. Arrillaga et al. (1987) reported that hypocotyl excised from *D. obscura* plantlets and cultured on media supplemented with NAA (1.0 mg/l) and BA (0.1 mg/l) or Kin (0.1 mg/l) + 8 g/l agar showed a good callus proliferation after 30 days in culture. By that time the calli were green and no embryos were observed. On the other hand, when hypocotyl explant were cultured on MS medium with 2,4-D + 2iP or coconut milk, soft whitish calli were produced after 30 days in culture. Cacho et al. (1991) obtained callus cultures when leaf, root, and hypocotyl explants from *D. thapsi* seedlings were cultured on MS medium with 2.0 mg/l 2,4-D. It was found that percentage of callus formation was 90 and 100 for leaf, root, and hypocotyl explants.

The effect of plant growth regulators and carbohydrate sources on callus induction, callus growth, in *D. lanata* was examined (Fatima et al. 2009). The leaf explant was very responsive and induced profuse calli, compared to other explants, such as stems, hypocotyls, and roots. Maximum callus growth was noted in MS supplemented with 6.0 mg/l of NAA and 3.0 mg/l of BA. Profuse callus induction was observed with NAA alone, but with cytokinins callus induction was further improved. It was found that maltose induced maximum callus biomass, as compared to other tested sugars. Maximum callus biomass (3.88 g fresh weight and 0.28 g dry weight) was obtained with 3% maltose after 12 weeks. Yücesan et al. (2014) showed that lamina explants from *D. lamarckii* seedlings developed green and nodular calli after 8 weeks of cultivation on (LS) medium containing 13.3 μM

Table 3.1 Studies on cell and callus cultures induction of *Digitalis* spp.

Explant used/species	Medium used	Culture condition	Growth regulators (mg/l)					Reference
			2,4-D	NAA	IAA	BA	Kin	
Anthers from flowering shoots/*D. purpurea*	W medium	In continuous light	10					Corduan and Spix (1975)
	NN	In continuous dark	5					
Leaf segment from 30-day-old seedlings/*D. obscura*	MS	Cultures were maintained 60 days at 25 °C with 16 h light period		1.0		2.0		Perez-Bermudez et al. (1984)
Hypocotyl from 20-day-old plantlets/*D. obscura*	MS	Cultures were incubated under 16-h photoperiod or darkness at 26 ± 2 °C for 30 days		1.0		0.1		Arrillaga et al. (1987)
Leaf, root, and hypocotyl explants/*D. thapsi* seedlings (30 days old)	MS	The cultures were kept in growth chambers at 26 ± 2 °C with a 16-h photoperiod	2.0					Cacho et al. (1991)
Leaves, roots, and hypocotyls were excised from germinated seedlings/*D. lanata*	MS	Cultures were incubated under a 16-h photoperiod at 25 ± 2 °C. The relative humidity was maintained at 60–70%		6.0		3.0		Fatima et al. (2009)
Lamina explants isolated from 4-week-old germinated seedlings/*D. lamarckii*	LS	The cultures were kept at 23 ± 1 °C under a 16-h light photoperiod at a relative humidity of 60%		10.8 μM		13.3 μM		Yücesan et al. (2014)
Root pieces or leaf from shoot cultures/*D. mariana*	MS	Cultures were cultivated in a growth chamber at 25 °C with a relative humidity of 60% and a 16-/8-h (light/dark) photoperiod	1.1 μM		5.7 μM		9.3 μM	Kreis et al. (2015)
Cotyledon and hypocotyl from in vitro seedlings/*D. davisiana*	MS	Cultures were incubated under 16-h light/8-h dark photoperiod	0.1					Verma et al. (2016)

W medium White (1943), *NN* Nitseh and Nitseh (1969), *MS* Murashige and Skoog (1962), *LS* Linsmaier and Skoog (1965)

BA and 10.8 μM NAA with 100% callus formation. Also, Kreis et al. (2015) obtained callus tissues from stock shoot culture (6-week-old) of seedlings *of D. mariana* when grown on MS medium with IBA 9.8 μM. Also, root pieces or leaf explants were excised from shoot cultures and grown on MS- medium supplemented with Kin (9.3 μM), IAA (5.7 μM), and 2,4-D (1.1 μM) produced a friable callus from the root explants. Recently, Verma et al. (2016) reported that MS medium with 0.1 mg/l 2,4-D was observed as the most effective and efficient medium, inducing 33 and 29.4% embryogenic calli, from cotyledon- and hypocotyl-derived callus, of *D. davisiana*, respectively. BAP was also effective at concentration of 0.5 mg/l for the two explants. Root explants did not response to all levels of 2,4-D or BAP in the culture medium.

3.2 Plant Regeneration of **Digitalis** *ssp.*

Propagation through tissue culture methods has provided an effective way of producing a large number of identical clones in a relatively short time as compared to conventional planting methods. Based on the plant species and culture conditions, in vitro propagation could be achieved by direct and/or indirect shoot organogenesis and/or somatic embryogenesis. Various *Digitalis* species cultivated in vitro showed clear differences in their regenerative behavior. All these regeneration methods have been successfully applied to Digitalis species and have been reviewed in depth. Studies on in vitro regeneration of *Digitalis* ssp. is presented in Table 3.2.

Corduan and Spix (1975) obtained regenerated plants from anther-derived callus of *D. purpurea.* They found that regeneration at high frequencies occurred only when the callus developed from anthers grown on the basal medium of NN-medium supplemented with 5.0 mg/l 2,4-D in continuous light or under certain photoperiodic conditions when kinetin at 2.2 mg/l was added to the basal medium. The regenerated plants were diploid (2n = 56 chromosomes) when the callus developed in the dark and tetraploid (4n = 112 chromosomes) when the callus was cultured in continuous light. Most of the regenerated plants and their progeny were euploid (having an exact multiple of the haploid number of chromosomes); they flowered and fruited normally and grew well under field conditions. Perez-Bermudez et al. (1984) reported that culturing leaf explants of *D. obscura* on MS medium with NAA (1.0 mg/l) + BA (2.0 mg/l) producing 30 buds per 15 developed shoots. Upon transfer of differentiated buds to medium without growth regulators, subsequently 50% of the transferred buds developed shoots. Root formation was achieved on MS medium with 0.5 mg/l IAA.

Hypocotyl-derived calli of *D. obscura* were obtained in agar-solidified medium with several growth regulator combinations giving rise to proembryo masses and globular embryos when transferred to liquid media with lower growth regulator and higher NH_4HO_3 levels (Arrillaga et al. 1987). They reported that, by transferring cultures from liquid media to different solidified media, new embryo formation took place. Somatic embryos showing normal cotyledonary stage developed

Table 3.2 Studies on in vitro regeneration of *Digitalis* ssp.

Species	Explant type	Establishment stage	Multiplication stage	Rooting stage	Reference
		Response			
D. purpurea	Anther-derived callus	NN; 2,4-D (5.0 mg/l) in continuous light NN; kinetin (2.2 mg/l) under photoperiodic conditions consisting of light at 28 °C and 8 h dark at 20 °C			Corduan and Spix (1975)
D. obscura	Leaf segment from 30-day-old seedlings	MS; NAA (1.0 mg/l) + BA (2.0 mg/l)	MS without growth regulators	MS; IAA (0.5 mg/l)	Perez-Bermudez et al. (1984)
D. obscura	Hypocotyl-derived calli	MS; NAA (0.2 mg/l) + BA (0.05 mg/l) (Somatic embryos)	MS; 2,4-D (0.2 mg/l) (New globular embryosarose)	MS; without growth regulators (Plantlets from embryos)	Arrillaga et al. (1987)
D. thapsi	Shoot tips from seedlings (30-day-old)	MS; NAA (0.5 mg/l) and Kin (0.1 or 0.5 mg/l)			Herrera et al. (1990)
D. thapsi	Leaf callus	Combinations of NAA with kinetin or BA were more effective in inducing organogenesis		MS; without growth regulators	Cacho et al. (1991)
	Root callus Hypocotyls callus	Using IAA in combination with BA gave the highest rate of shoot regeneration		IAA alone only induced the generation of roots	
D. obscura	Leaves from in vitro 30-day-old plantlets	MS; NAA (2.7 μM)	MS; free of growth regulators (Shoot elongation)	MS; free of growth regulators (Development of plantlets)	Vela et al. (1991)
	Axillary bud from matured plants	MS; BA (5 μM)	MS; free of growth regulators (Shoot elongation)	MS; free of growth regulators (Development of plantlets)	
D. lanata	Hairy root cultures	NmI; BA (8 μM/l) (Organogenesis)	NmI; without hormones	NmI solution; BA (15 μM/l) + NAA (125 μM/l)	Pradel et al. (1997)
	Hairy root cultures	liquid NmI; kinetin (0.1 μM/l) + 2,4-D (5 μM/l) (Somatic embryos)	NmI; containing BA (10 mM/l) + maltose (100 mM/l) in the light	NmI; maltose (100 mM/l) and without growth regulators	

D. purpurea	Hairy root cultures	1/2 MS; zeatin (0.5 mg/l)		1/2 MS (Complete regenerated plants)	Koga et al. (2000)
D. lanata	Callus from stems, leaves, roots, and hypocotyls	MS; BA (6 mg/l)	MS; Kin (1.5 mg/l) + IAA (0.5 mg/l) + Ch (500 mg/l)	MS; NAA (6 mg/l) + BA (3 mg/l)	Fatima et al. (2009)
D. lanata	Shoot tips	MS; NAA (0.5 mg/l)	MS; NAA (0.5 mg/l) + Kin (0.1 mg/l)	MS basal or MS; NAA (0.5 mg/l)	Ghanem et al. (2010)
D. trojana Ivan	Leaves	MS; NAA (0.1 mg/ml) + BAP (3.0 mg/ml)	MS; basal medium	MS; 0.1% (w/v) activated charcoal	Corduk and Aki (2010)
D. davisiana Heywood	Hypocotyl and flamingo-bill	LS; TDZ (0.5 mg/l) and IAA (0.25 mg/l)		LS; IAA (0.5 mg/l)	Gurel et al. (2011)
D. lamarckii Ivan	Leaves from in vitro germinated seedlings	MS; TDZ (1.0 mg/l)		MS; IAA (0.5 mg/l)	Verma et al. (2011b)
	Leaves from in vitro produced microshoots		MS; IBA (0.2 mg/l) + TDZ (0.2 mg/l)	MS; IAA (0.5 mg/l)	
D. lamarckii Ivan	Cotyledonary leaf from 3-week-old in vitro germinated seedlings	MS; NAA (0.54 μM) + BAP (2.22 μM) (Embryogenic callus)	MS; NAA (1.34 μM) + BAP (8.87 μM) (Somatic embryogenesis)	MS; IAA (2.5 μM)	Verma et al. (2011a)
		MS; NAA (2.69 μM) + BAP (8.87 μM) (Organogenic callus)	MS; NAA (2.69 μM) + BAP (8.87 μM) (Shoot organogenesis)	MS; IAA (2.5 μM)	
D. trojana Ivan	Hypocotyl segment from 21-day-old in vitro germinated seedlings	MS; TDZ (1.0 mg/l) + IAA (0.5 mg/l) (Initiation of somatic embryos)	1/2 MS; without growth regulator (Development of Plantlets)		Verma et al. (2012)
D. purpurea	Node, internode, and leaf	MS; BA (7.5 μM)	MS (Liquid); BA (7.5 μM)	MS; IAA (15 μM)	Patil et al. (2013)
D. lamarckii Ivanina	Leaf lamina (From 4-week-old germinated seedlings)	LS; BA (13.3 μM) + NAA (10.8 μM)	LS; GA_3 (1.4 μM) + BA (1.1 μM)	LS; IAA (2.8 μM)	Yücesan et al. (2014)

(continued)

Table 3.2 (continued)

Species	Explant type	Establishment stage	Multiplication stage	Rooting stage	Reference
		Response			
D. cariensis	Flamingo-bill	LS; TDZ (0.5 mg/l) + IAA (0.25 mg/l)	LS; TDZ (0.5 mg/l) + IAA (0.25 mg/l)	LS; IAA (1.0 mg /l)	Mohammed et al. (2015)
D. mariana	Shoot tips and lateral buds from seedlings	MS; IBA (9.8 μM) (Shoot cultures)		1/2 MS; IBA (9.8 μM) + agar (8 g/l) + 10 g/l sucrose	Kreis et al. (2015)
	Callus cultures	MS (Liquid); BAP (13.3 μM) + IAA (1.7 μM) (Adventitious shoots)		1/2 MS; IBA (9.8 μM) + agar (8 g/l) + 10 g/l sucrose	
D. ferruginea (subsp. *schischkinii*) (Ivanina)	Hypocotyls	MS; TDZ (2.0 mg/l) + IAA (0.5 mg/l)		MS; IAA (1.0 or 2.0 mg/l)	Yücesan et al. (2016)
D. davisiana	Cotyledonary leaf from in vitro germinated seeds	MS; BAP (1.0 mg/l) (Somatic embryos)	MS; NAA (0.25 mg/l) + BAP (3.0 mg/l) (Shoot regeneration)	NAA (1.0 mg/l)	(Verma et al. (2016)
D. purpurea cv. Berggold	Leaf segments from in vitro germinated seeds	MS; NAA (0.54 μM) and 6-BAP (13.20 μM)	MS; IAA (0.57 μM) + 6-BAP (4.4 μM)		Pérez-Alonso et al. (2018)
D. trojana Ivan	Hypocotyl segments	MS; IAA (0.5 mg/l)	MS; IAA (1.0 mg/l)	1/2 MS	Verma et al. (2018)
D. lanata (subsp. *Lanata*)	Shoot tips were excised from 3 week-old in vitro germinated seedlings	MS; IAA (0.25 mg/l)			Yücesan et al. (2018)

NmI (Lehmann et al. 1995), *MS* Murashige and Skoog (1962), *LS* Linsmaier and Skoog (1965), *TDZ* Thidiazuron

into vigorous whole plants when transferred to medium lacking growth regulators. In another study, shoot tips from seedlings of *D. thapsi* were cultured on MS medium with various auxins alone or in combination with cytokinins. The best results for shoot multiplication and direct rooting after 4 weeks of cultivation were obtained with 0.5 mg/l NAA and 0.1 or 0.5 mg/l kinetin. In that medium, the percentage of explant with shoot multiplication and formation of roots was 86 (Herrera et al. 1990).

The effects of the auxins 2,4-D, NAA, and IAA either alone or in combination with kinetin or BA were investigated to assess the morphogenetic potential of leaf, root, and hypocotyl explants of *D. thapsi.* (Cacho et al. 1991). The results showed that the presence of NAA induced root formation and that of BA induced shoot formation via callus interphase. IAA alone only induced the generation of roots in the hypocotyl callus. Kinetin was ineffective in all the explants tested. Combinations of NAA with kinetin or BA were more effective in inducing organogenesis in leaf explants. Optimum responses were obtained in hypocotyl and root explants by using IAA in combination with BA, the highest rate of shoot regeneration being observed in hypocotyl explants. Rooting of the differentiated shoots was readily achieved in media without growth regulators. The regenerated plantlets were transferred to soil and grew with a survival rate of 70%. Also, Vela et al. (1991) found that differentiation of somatic embryos and adventitious buds from cultured leaves of *D. obscura* required NAA alone or combined with BA, respectively. Somatic embryos were obtained on MS medium with 2.7 μM NAA, producing 5.1 embryos per leaf explant. Shoot proliferation from axillary buds was not affected by seasonal fluctuations in the stock plants and increased relative to the cytokinin concentration, but auxin reduced the multiplication rate. The highest number of shoots was obtained on MS medium with 5 μM BA producing 36.1 shoots per axillary buds explants.

Hairy roots obtained after transformation with *A. rhizogenesis* strain A4 were placed on solidified NmI-medium (Lehmann et al. (1995) supplemented with 8 μM/l BA and cultivated for 3–4 months in the light for regeneration (Pradel et al. 1997). After about 6 months, green spots appeared on several roots. They formed a multitude of more or less vitrified shoots during further cultivation. Rooting was initiated by bathing the shoots in a solution containing 15 μM/l BA and 125 μM/l NAA. The shoots were then cultivated on NmI solidified with perlite. Regeneration through somatic embryogenesis was achieved when pieces of hairy roots were cultivated on solidified NmI-medium containing 0.1 μM/l kinetin and 5 μM/l 2,4-D for about 3 months for callus formation. Suspension cultures were obtained from the callus in liquid NmI supplemented with 0.1 μM/l kinetin and 5 μM/l 2,4-D. Embryogenesis was triggered in NmI containing 100 mM/l maltose instead of sucrose and 10 mM/l BA in the light. After 10 weeks the embryos were spread on solidified NmI-medium without growth regulators and containing 100 mM/l maltose. Plants with 4–5 leaves were transferred to perlite. All steps were carried out upon continuous illumination. About 10% of the embryos formed plants with a well-developed root system and normal leaves. About 20% of the embryos developed into abnormal plants. About 70% of the embryos died.

Transformed root pieces of *D. purpurea* grown on 1/2 MS medium with 0.5 mg/l zeatin produced calli (Koga et al. 2000). Upon transfer to fresh medium, shoots were produced. The highest regeneration percentage was 16 on the previous medium. The shoots were cultured by means of the paper wick method with 1/2 MS liquid media. As a result of this treatment, a large number of healthy shoots were produced. In potted plants, values of plant height, number of leaves, spike length, number of florets, and leaf size were lower in a strain of transformants than in non-transformants. Fatima et al. (2009) showed that the regeneration efficiency was higher in leaf calli of *D. lanata* than in calli from other sources. The 6.0 mg/l BA concentration was more effective for the emergence of shoots. The maximum number of shoots increased with time and reached 133 shoots within 12 weeks of culture. Maximum growth was obtained on medium supplemented with 1.5 mg/l of kinetin +0.5 mg/l of IAA +500 mg/l of Ch. Glucose also showed good regeneration ability and produced 72.40 microshoots within 12 weeks of culture. For rooting, the frequency of rooting was maximal in MS + 6.0 mg/l of NAA + 3.0 mg/l of BA, and the average number of roots was 11.20. Sucrose at 3% promoted the rooting process to the highest degree, as both the number of roots and their growth improved.

Establishment of regeneration protocol from shoot tip explants of *D. lanata* germinated seedlings was studied (Ghanem et al. 2010). Results obtained revealed that MS medium containing 0.5 mg/l NAA was the best medium for shoot initiation. MS medium containing 0.5 mg/l NAA + 0.1 mg/l Kin was the best for multiplication of shoots where the number of leaves was 8.0 and the number of nodes was 16.00. The established shoot with root can be achieved directly in basal medium or MS medium supplemented with 0.5 mg/l NAA. The obtained plantlets were removed from the culture vessels and taken to pots containing a mixture peat moss and perlite 1:1. The plantlets were incubated in greenhouse for one month. In another study Corduk and Aki (2010) found that adventitious shoots were formed on leaf explants of *D. trojana* within 3 weeks in culture. The results showed that the best shoot proliferation (28.0 shoot/explant) was observed among explants cultured on MS medium with 0.1 mg/ml NAA + 3.0 mg/ml BAP. Regenerated shoots were multiplicated by subculture. For rooting, shoots were cultured on MS with 0.1% (w/v) activated charcoal. All of the in vitro regenerated plantlets were successfully acclimatized ex vitro and then grown healthy. Gurel et al. (2011) tested six different types of culture medium to assess the best medium for shoot initiation of *D. davisiana*. The results revealed that LS medium was the most effective for shoot production. Shoot regeneration efficiency was higher when flamingo-bill or hypocotyl explants were cultured on LS medium containing 0.5 mg/l (TDZ) and 0.25 mg/l (IAA). The highest number of shoots/explant (8.1) was recorded with flamingo-bill in that medium. Rooting of shoots that regenerated was readily achieved, and the highest number of root per shoot was 8.1 on medium with 0.5 mg/l IAA or even in the absence of plant growth regulators.

Verma et al. (2011b) found that, for shoot regeneration of leaf explants of *D. lamarckii*, TDZ was the most effective at 1.0 mg/l, producing a mean of 10.3 shoots per explant and was significantly more effective than BAP. Another experiment using leaf explants from in vitro produced shoots, it was found that a combination of 0.2 mg/l IBA with 0.2 mg/l TDZ produced significantly more shoots per explant (16.5 shoots) than with BAP (11.0 shoots), zeatin (5.5 shoots), or kinetin (4.0 shoots). The regenerated shoots were readily rooted on medium containing 0.5 mg/l (IAA). Rooted regenerants were then transferred to the pots, where they grew well and attained maturity. Also, Verma et al. (2011a) induced embryogenic callus on (MS) medium containing 0.54 μM (NAA) and 2.22 μM (BAP). Somatic embryos developed readily when the embryogenic callus was transferred to MS medium containing 1.34 μM NAA and 8.87 μM BAP, which produced a mean of 37.0 embryos per cotyledonary leaf explant. An organogenic callus was induced on MS medium containing 2.69 μM NAA and 2.22 μM BAP. Shoot development was observed when the organogenic callus was transferred to MS medium containing 8.87 μM BAP and 2.69 μM NAA. The highest mean number of shoots (5.67 shoots per explant) was obtained in that medium. The regenerated shoots were readily rooted on MS medium containing 1.0, 2.5, or 5 μM (IAA) or (IBA). Rooted regenerants were then transferred to the pots, where they grew well and attained maturity. Over 90% of the regenerants survived the hardening process.

Verma et al. (2012) reported that TDZ alone was found the most effective at 1.0 mg/l concentration, producing a mean of 10.7 somatic embryos per hypocotyl explant of *D. trojana*. Another experiment using MS medium with TDZ combined with different concentrations NAA, IAA, or IBA. A combination of 1.0 mg/l TDZ with 0.5 mg/l IAA produced significantly more somatic embryos per explant (13.8 embryos) than with BAP (8.8 embryos). During subculture on growth regulator-free half-strength MS medium, somatic embryos gradually developed into plantlets. Regenerated plantlets were successfully transplanted and grown in a greenhouse environment. In *D. purpurea*, Patil et al. (2013) described a protocol for in vitro propagation via direct organogenesis using several explants (nodal, internodal, and leaf). The highest response for shoot induction (85.7%, 12.7 shoots per explant) was obtained from nodal explants cultured on MS medium supplemented with 7.5 μM 6-BAP. To further maintain the multiple shoot induction, mother tissue was subcultured on fresh shoot induction liquid medium after each harvest. An average of 18 shoots per explant could be produced. Shoots were rooted in vitro on MS containing 15 μM IAA. Rooted plantlets were transferred to pots containing sand: soil mixture (1:1). Plantlets were maintained in a glass house.

Lamina explants of *D. lamarckii* seedlings developed callus when cultured on LS medium containing 13.3 μM BA and 10.8 μM NAA (Yücesan et al. 2014). Indirect shoot regeneration was achieved when the callus explants were cultured on LS medium supplemented with GA_3 (1.4 μM) + BA (1.1 μM) where 6.6 shoots/callus were produced. Regenerated shoots were rooted in LS medium with 2.8 μM IAA. The highest number of root/shoot (4.4) was recoded with that medium, while the highest root length (8.6) was observed when culture medium contained 1.4 μM IAA. About 95% of the regenerated plantlets were successfully acclimatized in a mixture of soil, manure, moss, and sand at a ratio of 1:2:2:1 (1,1, v/v) under nonaxenic conditions in the greenhouse. Mohammed et al. (2015) found that the highest mean number of shoots was produced on LS medium supplemented with a combination of 0.5 mg/l TDZ and 0.25 mg/l IAA, producing 3.9 shoots per flamingo-bill-type explants of *D. cariensis*. Regenerated shoots were rooted on LS medium containing 1.0 mg/l IAA, The greatest number of roots produced was 5.6 roots per shoot after 3 weeks of culture. Rooted plantlets were subsequently transferred to plastic pots filled with autoclaved commercial compost containing a mixture of soil, manure, moss, and sand at a 1:2:2:1 (w/w/w/w) ratio for hardening under greenhouse conditions with temperature varying between 20 and 25 °C and low relative humidity (20–35%). In another study, plants regenerated from either callus or shoot cultures of *D. mariana* were obtained through direct or indirect in vitro propagation (Kreis, et al., 2015). Multiple shoot formation was induced from a seed grown on agar-solidified (MS)-based medium supplemented with 9.8 μM (IBA). Shoots were used to regenerate plants directly, and some were then used as a source of root explants for callus induction. Callus was developed adventitious shoots when placed on MS medium containing BAP (13.3 μM) and IAA (1.7 μM), producing 5.8 shoot/explant. Developing shoots were isolated and transferred onto rooting media consisting of agar (8 g/l) half-strength MS medium with 9.8 μM IBA. Plantlets were transferred to pots containing sterile potting mix with 20% (w/w) sand and were transferred to the greenhouse.

Yücesan et al. (2016) studied in vitro regeneration using cotyledonary leaves and hypocotyl segments of *D. ferruginea*. They found that the most productive type of explant was hypocotyl segments. For shoot regeneration, the highest frequency of shoot organogenesis was obtained from MS medium containing 2.0 mg/l TDZ and 0.5 mg/l IAA, producing 11.0 shoots per explant with 100% regeneration frequency from hypocotyl explants after 4 weeks. For rooting, IAA was the most effective at 1.0 or 2.0 mg/l, producing 16.3 or 17.3 roots per shoot, respectively, after 4 weeks. For hardening off, upon transfer to pots containing a compost mixture, almost all regenerants displayed normal growth and morphological similarities to source plants in Trabzon, and they were maintained in a greenhouse for 6 months (Fig. 3.3.). However, plant regeneration through somatic embryogenesis was obtained from the cotyledonary leaf segments of *D. davisiana* (Verma et al. 2016). Maximum number of somatic embryos were achieved on MS medium supplemented with 1.0 mg/l BAP which produced 8.3 embryos per explant. Shoot regeneration was higher (14.3 shoots) in medium containing 0.25 mg/l NAA + 3.0 mg/l BAP. The shoots were successfully rooted on MS medium supplemented with NAA (1.0 mg/l) producing 5.3 roots per shoot. Regenerated plants were transferred to pots containing mixture of soil, manure, moss, and sand at ratio 1:2:2:1 and kept at 20–22 °C. The produced plants were successfully acclimatized and transferred to soil.

Recently, Pérez-Alonso et al. (2018) described a method to obtain in vitro plants of *D. purpurea* from leaf segments through direct organogenesis. The highest frequency of shoot regeneration (98.5%) with an average number of shoots per leaf segment of 18.9 was achieved via direct organogenesis from leaf segment on MS medium containing 0.54 μM NAA + 13.2 μM 6-BAP. De novo regenerated shoots spontaneously developed roots during the multiplication phase on MS medium supplemented with 0.57 μM IAA + 4.4 μM 6-BAP. Additionally, Random Amplified Polymorphic DNA (RAPD) analysis showed 100% monomorphic bands, which indicated genetic stability of the obtained plants. The de novo regenerated plantlets from the best treatment showed no significant morphological

Fig. 3.3 Flowering stage of *D. ferruginea* subsp. Schischkinii located in Trabzon district (**a**), effects of 2.0 mg/l thidiazuron (TDZ) and 0.5 mg/l indole-3-acetic acid (IAA) on shoot regeneration in cotyledonary leaf explants after 3 weeks of culture (**b**, **c**), shoots developing from hypocotyl explants after 5 weeks of culture (**d**), rooting of shoots on MS medium containing 1.0 mg/l IAA after 4 weeks of culture (**e**), and acclimatization of regenerants (8 months old) under greenhouse conditions (**f**). Bars = 15 cm (**a**); 0.8 cm (**b**); 0.2 cm (**c**); 1 cm (**d**); 1 cm (**e**); 2.5 cm (**f**). (Adapted from Yücesan et al. 2016)

differences compared to mother plants. More recently, Verma et al. (2018) reported that, IAA alone was found to be the most effective, at a concentration of 0.5 mg/l, inducing ~ 10 somatic embryos (SEs) per hypocotyl explant of *D. trojana* with 52% induction frequency. After 4 weeks of culture, the SEs initiated organogenesis and developed into complete plantlets (with 4–5 leaves). Also, they concluded that in the presence of an auxin polar transport inhibitor (such as TIBA), SE formation from the embryogenic cells was suppressed, resulting in the development of abnormal embryos. TIBA impeded the adventitious root formation as well. Also, an efficient shoot tip cultivation of *D. lanata* with one-step regeneration system for 12 weeks without applying an additional rooting stage was established by Yücesan et al. (2018). They demonstrated that the best regeneration medium was MS with 0.25 mg/l IAA or 0.50 mg/l IBA. The highest number of shoot/explant (2.0), number of root/explant (6.4) and root length (8.1) with 95% survival rate was recorded with MS medium supplemented with 0.25 mg/l IAA. The produced plantlets were transferred to plastic pots filled with autoclaved commercial compost containing a mixture of soil, manure, and sand at a 1:2:1 (w/w/w) ratio for 4 weeks. All the potted plantlets were covered with transparent lid to maintain high humidity for 1 week, and then all were moved to the greenhouse at 25 °C with a low relative humidity (20–35%).

3.3 *Cryopreservation of* Digitalis *ssp.*

It is well known that continuous subcultivation of cell cultures is expensive and time-consuming, and it carries the risk of reduction and loss the capacity of cell to accumulate secondary metabolites, besides somaclonal variation can be occurred. Thus, the necessity has arisen for efficient storage methods. In order to maintain secondary metabolite synthesis for long term, several cryopreservation protocols have been employed. The cryopreservation protocol comprises the following steps: preculture in media supplemented with osmotically active compounds, treatment with cryoprotective agents, slow freezing and storage at −196 °C, rapid thawing, post-thaw treatments, and recovery growth. In this respect, preculture is found to enhance the freeze tolerance of the cells markedly (Goldner et al. 1991). Cryogenic storage is a useful method for preserving various biochemical characteristics in plant cell cultures. *D. lanata* plants that have been found to produce cardenolides in large amounts are used for breeding and propagation programs (Seitz et al. 1983). The possibility of using in vitro techniques for conservation of elite plants is therefore of great value. The work of cryopreservation of in vitro cultures from *Digitalis* ssp. by various authors is summarized in Table 3.3.

D. lanata is one of the species where the biochemical capacities of frozen-thawed cells were studied very early. Diettrich et al. (1982) studied cryopreservation of *D. lanata* cell suspension culture. They reported that about 50% of the cells survived freezing and thawing. However, the apparently surviving cells showed signs of injury. The cultures derived from the surviving cells had the same growth rate and biochemical activity relative to the transformation of cardenolides, e.g., digitoxin, as the parent cultures. The frequency distribution of the nuclear DNA content in the cell cultures was the same before and after cryopreservation. These results indicate that there is no selection of a special cell type during freezing and thawing. Also, suspension cultures of *D. lanata* were stored at the temperature of liquid nitrogen (Seitz et al. 1983). They found that after thawing viable cultures were recovered which showed a growth rate equal to that of untreated cells. A comparison with the control curves revealed that no reduction of the biotransformation capacity occurred during the freeze-thaw procedure. On the contrary, they observed a minor increase of 10–20% in the maximum value of β-methyldigoxin produced by treated cells. The biotransformation capacity of cryopreserved cells in 300 ml shake cultures or in 20 l bioreactor had remained unchanged. Goldner et al. (1991) reported that, in *D. lanata* cell culture, the best results were obtained when trehalose or mannitol was used; the post-thaw viability values were 66% and 53%, respectively. Neither proline nor sorbitol was able to protect *Digitalis* cells sufficiently. In the range of 0.2–0.38 M trehalose, rapid growth was observed in the frozen-thawed cultures. Very high trehalose concentrations (0.89 M) caused deleterious effects. Cell viability showed a considerable decrease at the beginning of the preculture period. This reduction was found to be transient in the presence of trehalose, mannitol, sucrose, and melibiose. The phenylalanine ammonia lyase (PAL) activity increased markedly in the presence of proline, whereas the trehalose-treated and the control cells behaved nearly identical to one another. In another study, shoot tips

Table 3.3 Studies on cryopreservation of different cultures of Digitalis ssp.

Species	Culture type/ condition	Preculture additives	Cryoprotectant freezing/method Storage regime	Results; comments	Reference
D. lanata	Suspension cultures	Cell suspensions were grown on medium containing 3% mannitol	Cell suspensions were treated with a mixture of sucrose-glycerol (20%/20 V %), cooled slowly (about 1 °C/min) till −100 ° C and then were transferred to liquid nitrogen. After storage in liquid nitrogen, the cells were thawed rapidly in a water bath (40 °C) and spread on the surface of a solidified medium	About 50% of the cells survived freezing and thawing	Diettrich et al. (1982)
D. lanata	Cell suspension cultures (at 25 °C in the dark) on shake culture and 20 l airlift bioreactor	Mannitol (6%) was added on day 6 of the growth cycle to the culture medium After 3 days, the cells were subjected to the freezing procedure	DMSO, glycerol (0.5 M), and sucrose (1.0 M) were added to suspension medium. Cultures kept on shaker at +4 °C (I h.). The freezing regime, a cooling rate of I°C/ min, and then kept for 40 min at −35 °C followed by storage in liquid nitrogen. Samples were thawed in warm water (+40 °C) and then cultured in medium with 0.7% agar	Post-thaw viability of more than 50% of the cells was achieved The capacity of cells to transform *B*-methyldigitoxin to *B*-methyldigoxin remained unchanged after cryostorage	Seitz et al. (1983)
D. lanata	Cell suspension (strain 10/56-S50-Se) culture was cultivated on a gyratory shaker at 25 °C, in the dark	Trehalose, mannitol, sucrose, melibiose, proline, and sorbitol. On day 6 of the growth cycle, the culture medium was removed, and cells were transferred to the preculture medium for 3 days	DMSO, glycerol (0.5 M), and sucrose (1.0 M) Culture kept at 0 to +4 °C (I h.). The freezing regime, a cooling rate of I°C/ min, and then kept for 30 min at −40 °C followed by storage in liquid nitrogen (24 h). The samples were thawed in warm water (+40 °C) and then cultured in medium with 0.7% agar	Proline and sorbitol were not able to induce freeze tolerance. The best results with trehalose or mannitol The phenylalanine ammonia lyase (PAL) activity increased in the presence of proline, whereas the trehalose treated and the control cells behaved nearly identical to one another	Goldner et al. (1991)

(continued)

Table 3.3 (continued)

Species	Culture type/ condition	Preculture additives	Cryoprotectant freezing/method Storage regime	Results; comments	Reference
D. obscura	Shoot tips	Encapsulated shoot tips precultured 15 h in petri dishes containing MS with 0.4 M sucrose and 0.6% agar *Dehydration:* Shoot tips suspended in calcium-free MS with 0.1 M sucrose and 3% (w/v) sodium alginate and then dropped into MS with 0.1 M $CaCl_2$ and 0.1 M sucrose	Encapsulated shoot tips were placed in 2 ml cryotubes, directly plunged in LN and held there for 1 h Cryotubes were rapidly thawed in a water bath at 40 °C (2 min and the beads then transferred to petri dishes containing MS with 0.1 M sucrose and 0.6% agar to allow rehydration (24 h). Then, beads were transferred to MS medium with 1 M BA, 3% sucrose, and 0.8% agar m Beads were cultured for 30 days and regrowing buds excised and transferred to PM.	Prolonged subcultures reduced proliferation rates in both control and cryopreserved HU3 shoot tips, whereas long-term storage in LN did not affect the shoot recovery rate of the genotype Dehydration significantly decreased survival and shoot regeneration rate, but permitted the recovery of normal shoots after bead immersion in LN	Sales et al. (2001)

from several genotypes of the cardenolide-producing *D. obscura* were successfully cryopreserved using the encapsulation-dehydration technique (Sales et al. 2001). The results showed that precultivation on MS medium containing 0.5 M sucrose, followed by 2.5 h dehydration (final weight 30%), induced shoot regrowth in 42% of cryopreserved shoot tips. Encapsulated shoot tips showed a recovery rate of 100%. Shoot tips from the preculture with 0.5 M sucrose showed higher percentages of normal shoot development after 75 days. RAPD markers were employed to evaluate possible somaclonal variation occurring in shoots regenerated through successive subcultures and after cryopreservation. The band patterns revealed differences between the original parent plant and the shoots grown in vitro, especially after a prolonged subculture. Nevertheless, match percentages were higher (98.6% to 99.5%) when band patterns from subcultured shoots were compared to those obtained from their respective control or frozen progenies indicating that cryopreservation ensures genetic stability of selected genotypes.

3.4 Genetic Transformation of Digitalis ssp.

3.4.1 *Agrobacterium tumefaciens*-Mediated Transformation

Genetic transformation is a tool of special interest for developing new biotechnological strategies for the production of bioactive compounds such as cardenolides, which are exclusively obtained from plants. To date, *Digitalis* plants are the main economically viable source of cardenolides for the pharmaceutical industry (Perez-Alonso et al. 2014a). In spite of this, little information is available on genetic transformation of this economically important genus (Sales et al. 2003). Studies on *A. tumefaciens*-mediated transformation of *Digitalis* ssp. is presented in Table 3.4.

A repeatable transformation system has been established for *D. minor* using *Agrobacterium tumefaciens* (Sales et al. 2003). Leaf explants were inoculated with either EHA105 (carrying the nptII and gusA genes) or AGL1 (with the bar and gusA genes) strains. Among the tested factors influencing T-DNA transfer to plants, the EHA105 strain and the addition of acetosyringone to the co-culture medium increased transformation. The highest transformation efficiency (8.4%) was obtained when explants, soaked in a bacterial suspension with an OD_{550} of 0.9, were subcultured on selection medium after a 4-day co-culture with the bacteria. Evidence for stable transgene integration was obtained by PCR, Southern hybridization, growth on media selective for nptII or bar genes, and expression of the gusA gene. However, more than 200 transgenic plants were recovered from 65 independent explants. Thirty of these plants were successfully established in soil. Li et al. (2014) developed a rapid and efficient method for in vitro propagation and *A. tumefaciens*-mediated transformation of *D. purpurea*. Mature leaf explants were used for 100% adventitious shoot regeneration on (MS) medium supplemented with 1 mg/l thidiazuron (TDZ) + 0.1 mg/l (NAA). Transformation was achieved by

Table 3.4 *A. tumefaciens*-mediated transformation studies on *Digitalis* ssp.

Species	Infected explant	*Agrobacterium* strain	Response	Reference
D. minor	Leaf explants from 30-day-old seedlings	EHA105 harbors the p35SGUSINT plasmid (carrying the nptII and gusA genes) or AGL1 contains pTAB16 plasmid that includes the bar and gusA genes	Addition of 100 μM acetosyringone to the co-culture medium increased transformation AGL1 strain did not produce any transformed plants PCR analysis showed that the expected fragments for nptII (550 bp) and bar (507 bp) genes were found in all the analyzed plants Southern blot analyses confirmed the presence of the amplified fragments corresponded to the nptII, bar, and gusA genes	Sales et al. (2003)
D. purpurea	Leaf explants	GV2260/pBI121 or GV3101/pBI121	PCR and Southern blot analysis confirmed the presence of nptII and GUS on the transformed plants, and stable GUS expression was detected in the transformants by RT-PCR analysis	Li et al. (2014)
D. purpurea	Leaf explants	EHA105 and C58C1Rif^R (pMP90), both harboring the binary vector pTJK136	Transformation was confirmed by histochemical β-glucuronidase (GUS) assays of the putative transgenic tissues and PCR analyses using β-glucuronidase (uidA) and neomycin phosphotransferase II (nptII) Southern blot hybridization confirmed the stable integration of the nptII gene in the transgenic plants	Perez Alonso et al. (2014)

inoculating leaf explants with the *A. tumefaciens* strains GV2260/pBI121 or GV3101/pBI121. The binary vector pBI121 contained the reporter β-glucuronidase gene (GUS) and kanamycin selection marker nptII. Kanamycin-resistant shoots were regenerated directly on the selection medium 4–6 weeks after co-cultivation. Approximately, 52.2 and 60% of kanamycin-resistant shoots transformed with *Agrobacterium* strains GV2260 and GV3101, respectively, showed strong GUS staining by histochemical assay (Fig. 3.4). Recently, Perez-Alonso et al. (2014a) described the development of efficient plant regeneration and *Agrobacterium*-mediated genetic transformation protocols for *D. purpurea*. Callus induction was achieved when leaf explants were cultured on MS medium with 2,4-D (4.5 μM). For plant regeneration, the produced callus was grown on regeneration media. MS medium with 0.57 μM IAA and 4.4 μM BA was the best medium for regeneration. The percentage of regenerating callus (100%) and the number of regenerated shoots per segment (16.0) in that medium. Successful transformation was confirmed by histochemical, PCR analyses and Southern blot hybridization. Strain C58C1Rif^R

Fig. 3.4 Regeneration of kanamycin-resistant shoots on SIM medium and histochemical GUS assay. (**a**) kanamycin-resistant shoots emerging from leaf explants edge on SIM medium 4 weeks after inoculation, (**b**) histochemical GUS assay for kanamycin-resistant shoots after 6 weeks of culture on SIM medium, (**c**) root induction of transformants on RIM medium after 4 weeks of culture, (**d**) potted transgenic plant, (**e**) detection of GUS expression of a normal shoot from **c** (left) and mature leaf from **d** (right) by histochemical GUS assay. (Adapted from Li et al. 2014)

(pMP90) yielded a higher number of transformed plants than EHA105. It was found that in total, 518 independent transgenic lines were regenerated with an average of 6.91 transgenic lines per initial leaf segment infected with the strain C58C1Rif^R.

3.4.2 *Agrobacterium rhizogenesis*-Mediated Transformation

The use of *Agrobacterium rhizogenesis* has been receiving attention in secondary metabolites research. It inserts the Ri plasmid into wounded tissue, causing the growth of very fine adventitious roots, so-called "hairy root." These roots can be cultured in hormone-free medium, and there are several examples of enhanced accumulation of secondary metabolites, relative to non-transformed tissue. Studies in establishment of hairy root cultures from Digitalis *ssp.* have been attempted by few workers as shown in Table 3.5.

Table 3.5 Studies in hairy root culture from *Digitalis* ssp.

Species	Infected explant	*Agrobacterium* strain	Response	Reference
D. purpurea	Leaf discs (ca. 1 cm x 1 cm)	*A. rhizogenes* harboring both a wild agropine-type Ri plasmid, pRi15834, and a binary vector, pGSGlucl or pBI121, containing the chimeric genes	The chimeric *neo* and *gus* genes on a mini Ti vector are efficiently transferred into the genome of *D. purpurea* using a binary vector system based on a root inducing Ri plasmid, pRi15834 The production of cardioactive glycosides was positively correlated to the concentration of chlorophyll in the cells	Saito et al. (1990)
D. lanata	Leaves (pieces with a length >1 cm)	A4, 15834, TRI05, 11325, 13332 smooth, 13332 scalloped, 25818, RIOOO, R1022, R1200, tmr 338, and Melone 2–1 (Wild-type strains)	Formation of hairy roots occurred with strains A4, 15834, and TRI05 only Transformation was proven by detection of opines and by DNA hybridization with the probe pLJ 85	Pradel et al. (1997)
D. purpurea	Leaves (pieces about 5 mm square)	The mikimopine-type strain A13	1/2 MS filter paper medium is more suitable as a co-cultivation medium for the hairy root induction Southern blot analysis was confirmed in the integration of T-DNA in Ri plasmid of *Agrobacterium* into plant genome using the fragment in the *rolC* gene as a hybridization probe	Koga et al. (2000)

Genetic transformation of foxglove (*Digitalis purpurea*) by chimeric foreign genes and production of cardioactive glycosides was investigated (Saito et al. 1990). After 2–3 weeks of infection, very fine hairy roots appeared at the veins of the leaves. In the experiments using pGSGlucl, out of the 38 clones of excised roots, 10 clones grew on Km (50 mg/l) containing medium; of these 10 clones, 6 were positive for the opines; of these 6, 4 were positive for GUS enzymatic assay. The transgenic state of established transformed roots was confirmed by Southern blot analysis and by detection of agropine and mannopine. The results of ELISA indicated that the cardioactive glycosides were highly produced in the green transformed hairy roots. However, Pradel et al. (1997) infected leaves of axenically grown shoots and plants of *D. lanata* with different wild-type strains of *A. rhizogenes*. They found that only three strains (A4, 15,834 and TRI05) formed hairy roots. No cardenolides were detected in hairy roots and untransformed roots grown in the dark or in the light. However, anthraquinones and flavonoids were shown to occur. However, plants were regenerated from the hairy roots either via somatic embryos or shoots. Plants regenerated from hairy roots showed similar contents of cardenolides as untransformed shoots and plants. Roots of untransformed and transformed plants contained cardenolides that probably were formed in the shoots and transported into the roots. Also, Koga et al. (2000) showed that using a filter paper soaked with 1/2 MS liquid medium for co-cultivation of leaf pieces of *D. purpurea* with the *A. rhizogenes* resulted in inducing 30–40 hairy roots per leaf pieces. Most of the induced roots were preserving a green color. Four months after inoculation, the percentages of actively grown roots on the phytohormone-free 1/2 MS solid media were 94% in the case of using a 1/2 MS filter paper medium and 85% in the other case. Regeneration percentage (16) from the hairy root through callus in zeatin addition medium was achieved. Compared between potted plants and transformed plants, it was found that values of plant height, number of leaves, spike length, number of florets, and leaf size were lower in a strain of transformants than in non-transformants.

4 Cardenolide Production from In Vitro Culture of *Digitalis* ssp. Through Biotic, Abiotic Elicitation and Precursor Feeding

Elicitors at appropriate concentrations might act as signaling molecules which could be perceived by a receptor present on plasma membrane and thus initiate the signal transduction network involving regulation of gene expression responsible for biosynthesis of target compounds (Chavan et al. 2011).

4.1 *Abiotic Elicitation and Cardenolide Production from In Vitro Cultures of* Digitalis *ssp.*

4.1.1 Medium Composition and pH

The effects of type of medium composition and pH on accumulation of cardenolide production from different *Digitalis* cultures have been extensively investigated. Response of different in vitro cultures of Digitalis ssp. to medium compositions and pH is presented in Table 3.6.

Shoot-forming cultures of *D. purpurea* were grown in various modifications of MS medium (Hagimori et al. 1983). The results showed that digitoxin content was highest at one-third $CaCl_2$ and lowest at threefold $CaCl_2$. The digitoxin content was highest at threefold $MgS0_4$ (ca. 120% of the control) and decreased with both lower concentrations and at tenfold $MgS0_4$. Maximal digitoxin formation as well as maximal growth took place at the same concentration as in the basal medium. Growth was not affected in any case, whereas the digitoxin content was slightly increased by a threefold increase in the concentration of KI, $Na2MoO_4$, or $CuSO_4$ and slightly reduced by $ZnSO_4$. The optimal pH of the medium for digitoxin formation before

Table 3.6 Response of in vitro cultures of *Digitalis* to medium composition and pH

Species	Culture type	Abiotic elicitor	Response	Reference
D. purpurea	Cells of shoot-forming cultures	Reduction of the basal $CaCl_2$ concentration	Reduction of the basal $CaCl_2$ concentration by two-thirds promoted both digitoxin formation and shoot-differentiation	Hagimori et al. (1983)
		Effect of $MgS0_4$ Concentration	A threefold increase in the $MgS0_4$ concentration improved digitoxin formation	
		Minor elements (H_3B0_3, $MnSO_4$, $ZnSO_4$, KI, Na_2MoO_4, $CuSO_4$, and $CoC1_2$)	A threefold increase in the concentration of either KI, Na_zMo0_4, or $CuS0_4$ improved digitoxin formation.	
		$FeS0_4$ coupled with Na_2-EDTA	The optimal concentration of $FeS0_4$ coupled with Na_2-EDTA was similar to that in the MS basal medium.	
		Initial pH	The optimal pH of the medium before autoclaving being 6.	
D. lanata	Cell cultures of strain S-1	High $MnSO_4$ levels (Incubation in 12 h/12 h of light and darkness)	Growth, digitoxin and chlorophyll content were decreased	Ohlsson and Berglund (1989)
	Cell cultures of strain S-2	High $MnSO_4$ levels (Incubation in 24 h darkness)	Digitoxin content was increased in the beginning of the culture period (days 0 and 6)	
		High $MnSO_4$ levels with GA_3 (1 μM) in the culture medium	Maximum digitoxin content was 142% of the control (GA_3 1 μM)	
D. thapsi	Cell suspension cultures	Absence of $CaCl_2$, addition of $MnSO_4$ and LiCl	Elimination of $CaCl_2$ promoted digoxin formation, $MnSO_4$ and LiCl increased digoxin content	Corchete et al. (1991)
D. obscura	Vitrified plants and normal plants	Changes in the concentration of MS macronutrient	Cardenolide content was lower in vitrified plants than in normal cultures	Lapena et al. (1992)
D. thapsi	Cell suspension cultures from hypocotyl and leaf-derived callus	Removal of calcium ions from medium	Removal of calcium from the medium reduced the growth and increase cardenolide of both lines	Cacho et al. (1995)

(continued)

Table 3.6 (continued)

Species	Culture type	Abiotic elicitor	Response	Reference
D. thapsi	Cell suspension cultures from hypocotyl and leaf-derived callus	Calcium deprivation	The elimination of calcium favored cardenolide production independently of the origin of the suspensions and the light regime	Cacho et al. (1999)
D. davisiana Heywood, *D. lamarckii* Ivanina, *D. trojana* Ivanina, and *D. cariensis* Boiss.	Callus cultures of hypocotyl segment from 1-month old seedlings	Elimination of calcium (Ca), magnesium (Mg), or both from the medium	Higher amounts of five cardenolides from MS medium lacking both Ca and Mg	Sahin et al. (2013)

autoclaving was 6, whereas growth was not particularly affected. Also, elevation of the $MnSO_4$ concentration in culture medium from 0.1 to 10 mM influenced growth and cardenolide accumulation in tissue cultures of *D. lanata* (Ohlsson and Berglund 1989). In strain S-1 in light, the digitoxin content decreased from 305 to 235 μg/g d.w., and the chlorophyll content decreased from 0.70 to 0.46 mg/g d.w. However, the digitoxin content increased from 16.7 to 63.5 μg/g (d.w.) in strain S-2 in darkness. Addition of $MnSO_4$ to strain S-2 in the beginning of the culture period (days 0 and 6) resulted in higher cardenolide content than later addition (days 10 and 15). With gibberellic acid in the medium, there was only a small increase in cardenolide accumulation in strain S-2 when the $MnSO_4$ concentration was elevated. However, Corchete et al. (1991) reported that the elimination of calcium from cell suspension cultures of *D. thapsi* resulted in an increase in the production of cardenolides on day 6, the number of equivalents of digoxin was 145% with respect to the 100% observed in cells growing in complete medium on the 12th day. Under such conditions growth was not affected. The highest digoxin relative content (%) was 1188 and 900 with 1.0 mM $MnSO_4$ after 7 and 15, days respectively. The highest digoxin relative content (%) was 297 and 851 with 100 μM LiCl after 7 and 15 days, respectively.

Lapena et al. (1992) showed that cardenolide content of in vitro *D. obscura* regenerated plants was lower in vitrified plants cultured on modified MS medium (half-strength NH_4NO_3) with 0.57 μM IAA and 4.40 μM BA, than in normal cultures with an overall reduction of photosynthetic pigments, lignin, and dry matter. Digoxigenin equivalents were 7.92 and 4.84 μg/g dry weight for normal plants and vitrified plants, respectively. Cacho et al. (1995) reported that the removal of calcium ions from MS culture medium induced a marked increase in the accumulation of cardenolides in cell suspension cultures of *D. thapsi*. This increase was more pronounced when the suspensions grew less actively, i.e. at the beginning and at the end of the experimental period. Cell viability was not affected although growth was slightly reduced. The results also suggest that calcium may play a role in the regulation of cardenolide metabolism in a concentration-dependent manner.

The effect of calcium deprivation on growth and the production of cardenolides in two undifferentiated cell lines of *D. thapsi* derived from leaf and hypocotyl callus and maintained under three different light regimes (16 h photoperiod, darkness, or continuous light) was investigated (Cacho et al. 1999). Growth was stimulated by continuous light in both cell lines cultured in complete medium. The light regime did not affect cardenolide accumulation in the cells of the hypocotyl-derived line; by contrast, continuous light or darkness increased the production in the leaf-derived line. It was concluded that the elimination of calcium favored cardenolide production independently of the origin of the suspensions and the light regime, this beneficial effect being predominantly manifested under continuous light. Recently, Sahin et al. (2013) reported that higher amounts of five cardenolides and total cardenolides were obtained when callus cultures of four *Digitalis* species were incubated on MS medium lacking both Ca and Mg. The mean contents of total cardenolides obtained were in the order of *D. lamarckii* (2017.97 μg/g), *D. trojana* (1385.75 μg/g), *D. cariensis* (1038.65 μg/g), and *D. davisiana* (899.86 μg/g) when both Ca and Mg were eliminated from the medium, respectively.

4.1.2 Subculture Time, Explant Age, and Type

Different research work has been reported to explore the effect of subculture time, explant age, and type on the accumulation of cardenolides. Response of different in vitro cultures of Digitalis ssp. to subculture time, explant age, and type is presented in Table 3.7.

Table 3.7 Response of in vitro cultures of *Digitalis* to subculture time, explant age, and type

Species	Culture type	Abiotic elicitor	Response	Reference
D. lanata	Extracts of cell suspension derived from leaf and root and in vivo plant leaf and root extracts	Effect of culture age (Leaf organ cultures (2, 4, 6, 8, and 12 weeks) and root organ cultures (4 weeks). Leaves and roots were collected from 26-week-old plants)	Leaf cultures at 8 and 12 weeks old contained more Lanatoside A, B, C than 4-week-old leaf cultures	Lui and Staba (1981)
D. lanata Ehrh	Callus cultures (from different explant types)	Effect of subculture time	Increasing the time of subculture led to an increase in digoxin and digitoxin production	Bosila et al. (2003)
		Effect of explant type	Highest content of digoxin and digitoxin with hypocotyl-derived callus	
		Effect of explants age	The highest content of digoxin and digitoxin was achieved with callus derived from 20-week-old leaf explants	

Lui and Staba (1981) found that the cardenolide from leaf and root culture of *D. lanata* extracts were similar to those obtained from the mother plant. The cardenolide pattern of 4-week-old leaf and root culture extracts was similar to that obtained from plant leaf and root extracts. In addition, as the age of the leaf culture increased (8 and 12 weeks), many of the unknown compounds disappeared and primary glycosides formed. Young organ cultures (4-week-old) contained primarily digitoxigenin aglycone and that older cultures (8- and 12-week-old) contained primarily digoxigenin and gitoxigenin aglycones. However, Bosila et al. (2003) reported that the highest content of digoxin and digitoxin (1.05 and 2.42 μg/g d.w. respectively) was extracted from callus produced from hypocotyl explants of *D. lanata*. In addition, the highest value of digitoxin was from callus derived from 2-week-old seedlings (0.023 μg/explant), while the highest amount of digoxin was in callus obtained from 4-week-old seedlings (0.028 μg/explant). Moreover, the continuous reculturing of callus induced its glycosidal content, which reached its maximum in the third subculture period. The highest accumulation of digoxin or digitoxin was observed when callus was subcultured from leaf explants and repeated three times, which gave 0.058 and 0.67 μg/g d.w. for digoxin and digitoxin, respectively.

4.1.3 Growth Regulators

Different research work has been achieved to explore the effect of growth regulators on the accumulation of cardenolides. Response of in vitro cultures of *Digitalis* ssp. to different types and levels of growth regulators is presented in Table 3.7.

Lui and Staba (1981) found that the highest gibberellic acid (GA) level tested (10 ppm) increased digoxin production approximately 40% (4.0 mg% dry wt.) in root cultures of *D. lanata*. It was found that digoxin production was 42 mg % dry wt. in the leaf cultures and 4.1 mg % dry wt. in root culture when grown in medium containing 50 ppm mefluidide. On the other hand, digoxin production in leaf cultures increased 60% (52 mg% dry wt.) upon removing the growth regulator (BA) from the medium. Hagimori et al. (1982) showed that the light-grown, green, shoot-forming cultures of *D. purpurea* accumulated considerable amounts of digitoxin (about 20 to 40 μg/g dw), and the white, shoot-forming cultures without chloroplasts accumulated about one-third that amount of digitoxin. The chlorophyll content and the ribulose bisphosphate carboxylase activity of the undifferentiated green cells were about the same as they were in the green, shoot-forming cultures, but the digitoxin content of the former was extremely low (about 0.05 to 0.2 μg/g dw), which is about the same as that in undifferentiated white cells without chloroplasts. Ohlsson and Bjork (1988) reported that cell suspension of strain (S-1) of *D. lanata* grown in light condition showed increased growth but decreased cardenolide content after cultivation with GA_3 in the medium, while strain (S-2) in darkness showed increased growth and cardenolide accumulation after GA_3 addition. With 1 μM GA_3 the cardenolide yield was more than 400% of that of the control.

Ohlsson (1990) found that the negative effects of ABA on growth and cardenolide accumulation were eliminated by GA_3 (1.0 μM) in light-sensitive cell suspension of

D. lanata (strain S-2). Growth of strain S-l in darkness, which was stimulated by GA_3, was further increased when ABA was added. For strain S-l in light, ABA was negative for both growth and cardenolide accumulation, while GA_3 was positive for growth and negative for cardenolide accumulation. Tissue differentiation was stimulated by GA_3 in S-lL and S-2D. This stimulation was decreased when ABA was also added. The pH in the culture medium at harvest was lower in strain S-2 than in strain S-1. Berggren and Ohlsson (1991) found that the effect of ASN or GA_3 on cardenolide accumulation of two strains from *D. lanata* cultures was different depending on the strain and the culture conditions. In strain S-1 (incubated in light) ASN (50 mg/l) or GA_3 (1.0 μM) decreased cardenolide production, while in strain S-2 (incubated in dark) addition of SAN increased the yield of cardenolides. Moreover, Berglund and Ohlsson (1992) reported that cardenolide accumulation in two strains (S-1 and S-2) cell suspension cultures of *D. lanata* was increased by the ethylene biosynthesis inhibitor aminoethoxyvinylglycine (AVG) in concentration of 0.5 mg/l and decreased by the ethylene releasing substance ethephon (5 mg/l). The stimulatory effect of AVG was dependent on the stage of growth at which it was added to the culture. Cardenolide accumulation was greatly enhanced by a high medium content of $MnSO_4$. When grown in this medium, the cardenolide accumulation in the tissue was 440% of the control. The effect of ethephon was similar in the two strains; 5 mg/l resulted in a reduction of the cardenolide accumulation to 40% of the control in strain S-l and 20% of the control in strain S-2. Aawad and AlKhateeb (2006) showed that treated root cultures derived from shoot tips explants of *D. purpurea* with 0.5 mg/l of IBA had a high significant effect on the dry weight of roots formation and contents of digitoxin and gitoxin whose quantities as a rate of 2.96 g, 30.01, and 11.05 μg/g dw, respectively.

Perez-Alonso et al. (2012) reported that morphological response of shoots of *D. lanata* cultivated in temporary immersion systems and elicited with methyl jasmonate was influenced by the elicitor. A reduction in length and number of shoots was evident with all MJ concentrations. The highest content of digoxin was recorded with MJ (60 μM). Lanatoside C content was slightly increased by MJ; significant differences were only observed between the highest concentration (100 μM) and the control. Moreover, increased content of H_2O_2 and MDA was only found with the highest MJ concentration. Also, Patil et al. (2013) reported that analysis of the shoot cultures of *D. purpurea* showed maximum accumulation of digitoxin (53.6 μg/g DW) and digoxin (28.8 μg/g DW) 28 days after inoculation. However, incorporation of auxins in the medium promotes the cardiac glycoside accumulation. About two-fold increase in digitoxin and 1.8-fold increase in digoxin content was observed in the medium fortified with 5 μM IAA. While the increase in digitoxin was 1.7-fold, digoxin was augmented by 1.55-fold on medium containing 5 μM NAA. Pérez-Alonso et al. (2014a) reported that elicitation with methyl jasmonate resulted in decreased biomass production on shoots of *D. purpurea* cultured on semisolid media. Digoxin and digitoxin content was slight and significantly increased by methyl jasmonate 80 and 100 μM, respectively. Recently, Yücesan et al. (2016) obtained regenerated plants from cotyledonary leaves and hypocotyl segments *of D. ferruginea* grown in MS medium with different growth regulators. The results showed that there was no significant difference between the materials in terms of their cardenolide profiles, irrespective of the medium formulation on which the shoots were produced. Moreover, all the regenerants displayed cardenolide patterns similar to those of plants collected from natural resources (Trabzon district). The highest percentage (w/w) of cardenolide C was found in regenerants grown on MS medium with TDZ (2.0 mg/l) + IAA (0.5 mg/l), while the highest % of digoxin was observed on MS with kin (0.5 mg/l) + IAA (1.0 mg/l). On the other hand, MS with BA (1.0 mg/l) + IAA (0.5 mg/l) gave the highest % of digitoxin (Table 3.8).

4.1.4 Light and Dark

The role of light in cardenolide production of *Digitalis* cultures has been extensively studied by different workers. Response of in vitro cultures of *Digitalis* to light and dark (photoperiod) is presented in Table 3.9.

Influence of light on the accumulation of cardenolides in *D. lanata* was studied (Ohlsson et al. 1983). A medium containing 0.2 mg/l BA stimulated cardenolide accumulation by strain S-1 in the light and by strain S-2 in the dark. Optimal irradiance for chlorophyll accumulation, viability, and growth was obtained at 10.0 watts of white light/m^2. With increasing chlorophyll content, there was an increase in the digitoxin content. No correlation could be found between the digoxin and the chlorophyll content. The digoxin content did not vary within the examined light conditions. Light from the yellow region had little or no effect. However, SAN 9789 inhibited cardenolide formation in the light but stimulated its accumulation in darkness. Ohlsson and Bjork (1988) found that cell suspension of strain (S-2) from *D. lanata* grown in darkness showed an increase in both growth and cardenolide accumulation after GA_3 addition. With 1 µM GA_3, the cardenolide yield was more than 400% of that of the control. Cacho et al. (1999) found that continuous light promoted growth in two undifferentiated cell lines of *D. thapsi*. However, no significant differences were observed in the cardenolide content under any of the three light regimes studied. The light regime did not affect cardenolide accumulation in the cells of the hypocotyl-derived line by contrast; continuous light or darkness increased the production in the leaf-derived line. The elimination of calcium from the culture medium reduced growth and significantly increased cardenolide production in the three light regimes. However, it was reported that shoot cultures of the cardenolide-producing species *D. lanata* accumulated up to 0.6 µM/g dw cardenolides when cultivated under continuous white light. After transfer to permanent dark, the cardenolide content of cultured shoots gradually decreased and reached non-detectable levels after 12 weeks. After transfer back to light conditions, cardenolides started to accumulate and reached the levels of light-grown controls after 4 weeks (Eisenbeiss et al. 1999). In another study, Bosila et al. (2003) found that the optimal callus growth of *D. lanata* was obtained at 16 h. light/day but the best glycosidal content was achieved when callus was exposed to 18 h. light/day. However, digoxin formation in callus negatively responded when increasing the length of the photoperiod up to 16 h. light/day. The highest content of digoxin (0.032 µg/g d.w.) was formed in callus which had been subjected to the shortest photoperiod of 10 h. light/day. On the other hand, a linear increase in digitoxin content was correlated with increasing the photoperiod, reaching a maximum with 18 h. light/day.

Table 3.8 Response of in vitro cultures of *Digitalis* to growth regulators

Species	Culture type	Abiotic elicitor	Response	Reference
D. lanata	Extracts of cell suspension derived from leaf and root and in vivo plant leaf and root extracts	Effect of growth regulators (IAA, BA, GA, and MF	GA (10 ppm) increased digoxin production (40%) in root cultures GA (5 ppm) increased digoxin production (34%) in leaf cultures MF (50 ppm) increased digoxin production (30%) in the leaf cultures and 40% in root culture	Lui and Staba (1981)
D. purpurea	Cell cultures and shoot-forming cultures	Effects of IAA, BA, NAA, and 2,4-D concentrations	The optimum concentrations of the tested compounds for accumulation of digitoxin were BA, 0.01 to 1.0 mg/l; IAA, 0.1 to 1.0 mg/l; NAA 0.1 mg/l; and 2,4-D, 0.01 mg/l	Hagimori et al. (1982)
D. lanata	Cell suspension (strain S-1)	Effect of GA_3	Showed increased growth but decreased cardenolide content after cultivation with GA_3 in the medium	Ohlsson and Bjork (1988)
	Cell suspension (strain S-2)		Showed increased growth and cardenolide content after cultivation with GA_3 in the medium	
D. lanata	Cell suspension (strain S-1) grown in light or darkness	Effect ABA and/or GA_3 both at Conc. 1.0 μM	*In light*, ABA had a negative effect on growth and cardenolide content; the effect of GA_3 was positive for growth and negative for cardenolide content *In darkness*, which was stimulated by GA_3, growth was increased by adding ABA	Ohlsson (1990)
	Cell suspension (strain S-2) grown in darkness		The negative effects of ABA on growth and cardenolide accumulation were eliminated by GA_3	
D. lanata	Cell suspension (strain S-1) grown in darkness	GA_3 concentration = 1.0 μM SAN concentration = 50 mg/l	In strain S-1 (light) ASN or GA_3 decreased cardenolide production	Berggren and Ohlsson (1991)
	Cell suspension (strain S-2) grown in darkness		In strain S-2 (dark) addition of SAN increased the yield of cardenolides	

D. lanata	Cell suspension (strain S-1) grown in light or darkness	Effect of ethephon[a] (5 mg/l)	Reduction digitoxin content to 40% of the control	Berglund and Ohlsson (1992)
		Effect of AVG (0.5 mg/l)	Increased digitoxin content by early addition of AVG	
	Cell suspension (strain S-2) grown in darkness	Effect of ethephon[a] (5 mg/l)	Reduction digitoxin content to 20% of the control	
		Effect of AVG (0.5 mg/l)	Increased digitoxin content by a high medium content of $MnSO_4$.	
D. purpurea (Var. Excelsior Mixed)	Root culture from shoot tips explants	Effect of different concentrations of IAA, IBA, and NAA.	Auxins concentrations (IBA) had a positive and significant effect on the production of digitoxin and digoxin than control treatment	Aawad and AlKhateeb (2006)
D. lanata	Shoots cultured in temporary immersion systems	Methyl jasmonate MJ (60, 80 and 100 µM)	The highest content of digoxin and Lanatoside C was found with 60 and 100 µM MJ respectively 100 µM MJ increased content of H_2O_2 and MDA	Perez-Alonso et al. (2012)
D. purpurea	Shoot cultures	MS medium containing different concentrations (0 to 15 µM) and combinations of cytokinins (BA, Kin, and TDZ) and auxins (IAA, NAA, and 2,4-D)	Cultures treated with MS medium fortified with 7.5 µM BA were found better for shoot growth and accumulation of cardiotonic glycosides	Patil et al. (2013)
D. purpurea (var. Roter Berggold)	Shoot cultures	Methyl jasmonate MJ (60, 80 and 100 µM)	Digoxin and digitoxin content was significantly increased by methyl jasmonate 80 and 100 µM, respectively	Pérez-Alonso et al. (2014a)

(continued)

Table 3.8 (continued)

Species	Culture type	Abiotic elicitor	Response	Reference
D. ferruginea (subsp. *schischkinii*)	Leaf samples of 8-month-old regenerants	MS; TDZ (2.0 mg/l + IAA (0.5 mg/l) MS; kin (0.5 mg/l) + IAA (1.0 mg/l) MS; BA (1.0 mg/l) + IAA (0.5 mg/l)) were used to determine the effects of plant growth regulators on cardenolide production	All treatments displayed similar cardenolide contents, i.e., the lanatoside C level was around 0.3% (w/w), while the levels of lanatosides A and B, digoxin, and digitoxin were less than 0.08% (w/w), similar to plants grown from seeds and collected from natural	Yücesan et al. (2016)

MDA malondialdehyde
SAN SAN 9789 4-chloro-5-(methylamino)-2-(α, α, α,-trifluoro-m-tolyl)-3(2H)-pyridazinone
AVG aminoethoxyvinylglycine
ABA abscisic acid
MF mefluidide
[a]Ethephon is the most widely used plant growth regulator. Upon metabolism by the plant, it is converted into ethylene, a potent regulator of plant growth and ripeness

Table 3.9 Response of in vitro cultures of *Digitalis* to light and dark (photoperiod)

Species	Culture type	Abiotic elicitor	Response	Reference
D. lanata	Cell suspension derived from anther filaments and kept on a rotatory shaker at 175 rpm and 12 + 12 h light and dark periods (Strain-1)	Incubation in light Values for the irradiation in the visible region are 10.0 W/m^2 for white, 1.3 W/m^2 for blue, 2.9 W/m^2 for green, 2.8 W/m' for yellow, and 2.1 W/m^2 for red light	The highest cardenolide and chlorophyll contents were obtained using light in the blue region	Ohlsson et al. (1983)
	(Strain-2)	Incubation in dark	Increase in cardenolide production when cultured on medium with BA but not in medium with 2,4-D	
D. lanata	Cell suspension (strain S-1)	Effect light and darkness	This strain accumulated cardenolides predominantly in light	Ohlsson and Bjork (1988)
	Cell suspension (strain S-2)		This strain in darkness showed an increase in both growth and cardenolide accumulation after GA_3 addition	
D. thapsi	Cell suspension cultures from hypocotyl and leaf-derived callus	Different light regimes (16 h photoperiod, darkness or continuous light)	The light regime did not affect cardenolide accumulation in the cells of the hypocotyls-derived line. Continuous light or darkness increased the production in the leaf-derived line	Cacho et al. (1999)
D. lanata	Shoot cultures from leaf meristems of axillary buds	Shoots were grown either under continuous light or in the dark on a rotary shaker (80 rpm and 20 ± 2 °C)	Shoot cultivated under continuous white light when transferred to permanent dark, the cardenolide content gradually decreased Transfer back to light conditions, cardenolides started to accumulate	Eisenbeiss et al. (1999)
D. lanata Ehrh	Leaf-derived callus cultures	Effect of photoperiods (10 h light/14 h dark cycle, 14 h light/10 h dark cycle, 16 h light/8 h dark cycle, and 18 h light/ 6 h dark cycle)	The highest content of digoxin was formed in callus which had been subjected to the shortest photoperiod of 10 h. light/day The highest content of digitoxin was with 18 h light/day	Bosila et al. (2003)

4.1.5 Chemicals

Response of in vitro cultures of *Digitalis* to chemicals is presented in Table 3.10. Cacho et al. (1995) showed that, addition of strontium, the growth of *D. thapsi* cells was significantly reduced. Its effect on cardenolide accumulation was similar to calcium. In the presence of EGTA, the accumulation of cardenolides increased in a concentration-dependent manner. Lanthanum also exerted a promoting effect on cardenolide accumulation in a concentration- and time-dependent manner. Also, Ghanem et al. (2010) showed that the total cardiac glycosides of regenerated plantlets of *D. lanata* were significantly decreased by addition of salicylic acid to the culture medium except at concentration of 200 μM where cardiac glycosides content was 0.707 mg/g D.W, which showed no differences with content of the control (0.759 mg/g D.W). The cardiac glycosides were also increased by addition $CaCl_2$ to the culture medium. The highest value of cardiac glycosides (2.204 mg/g DW) was found when 200 mM of $CaCl_2$ were added to the culture medium. Bota and Deliu (2011) investigated the effect of abiotic elicitor copper sulfate on the production of flavonoids in cell cultures of *D. lanata*. Two cell lines were used (line 11 and line 13C-100), and two types of experiments were performed. In the first type of experiment, the highest production of flavonoids was established, for both cell lines, for the strongest elicitor concentration (8.0 μM). In the second type of experiment, the maximal flavonoid production was induced for line 11 after a 24-hour elicitation (over 10 times more compared with the control, from 0.624 mg/g d.w. to 6.0 mg/g d.w.), for the highest elicitor concentration (40 μM). Moreover, Patil et al. (2013) reported that salicylic acid at 200 μM improved the accumulation of digoxin by 7.5-fold in shoot cultures of *D. purpurea*. Presence of 200 and 300 mM mannitol in the medium resulted in 4.8-fold and 2.57-fold higher accumulation of digoxin and digitoxin in the shoot biomass, respectively. Supplementation of 100 and 200 mM sorbitol was found optimum for 7.48- and 3.45-fold higher accumulation of digoxin and digitoxin, respectively. Enhanced digoxin accumulation was noticed with increasing concentrations of PEG-6000. About threefold higher digoxin accumulation was recorded at 5 mM PEG-6000 over that of control. About 8.7-fold higher digoxin accumulation was noticed in the shoot cultures grown on 200 mM KCl. NaCl was found ineffective in stimulating the synthesis of cardiotonic glycosides.

4.2 *Biotic Elicitation and Cardenolide Production from In Vitro Cultures of* Digitalis *ssp.*

4.2.1 Yeast Extract and Chitin

Many authors have mainly used yeast extract and chitin as biotic elicitors. Response of in vitro cultures of *Digitalis* to yeast extract and chitin is presented in Table 3.11.

Table 3.10 Response of in vitro cultures of *Digitalis* to chemicals

Species	Culture type	Abiotic elicitor	Response	Reference
D. thapsi	Suspension cultures from 4-month-old hypocotyl-derived calluses	Effect of strontium ($SrCl_2$), EGTA and lanthanum ($LaCl_3$)	In the presence of EGTA (10 μM), La^{3+} (0.28 μM), and Sr^{++} (3 mM), accumulation of cardenolides increased (depend on age of culture)	Cacho et al. (1995)
D. lanata	In vitro produced plantlets	Salicylic acid (0, 50, 100, 150, and 200 μM) and KCl (50, 100, 150, 200 mM)	The highest cardiac glycoside content was observed with 200 μM salicylic acid and 200 mM KCl	Ghanem et al. (2010)
D. lanata	Cell suspension cultures (cell lines 11 and 13C -100)	Copper sulfate Exp. 1: 1 ml $CuSO_4$ was added in flasks containing 0 days suspension cultures to obtain a final copper concentration of 2, 4, and 8 μM Exp. 2: $CuSO_4$ was added in flasks containing 10 days suspension cultures to obtain a final copper concentration of 20 and 40 μM	Increasing of flavonoid production for all elicitor concentrations and for both cell lines. Copper concentration (40 μM) accumulated the greatest flavonoid content after 24 h	Bota and Deliu (2011)
D. purpurea	Shoot cultures	Various concentrations of salicylic acid (0, 50, 100, 150, 200, and 250 μM), mannitol and sorbitol (0, 100, 200, 300, 400, and 500 mM), PEG-6000 (0, 1, 2, 3, 4, and 5 mM), NaCl (0, 20, 40, 60, 80, and 100 mM), and KCl (0, 40, 80, 120, 160, and 200 mM)	Increasing concentrations of salicylic acid, mannitol, sorbitol, PEG-6000, NaCl, and KCl reduced the growth but resulted in enhanced accumulation of cardiac glycosides	Patil et al. (2013)

EGTA ethylene glycol-bis(β-aminoethyl ether)-N,N,N′,N′-tetraacetic acid
KCl potassium chloride
PEG polyethylene glycol
NaCl sodium chloride
$LaCl_3$ lanthanum
LiCl lithium chloride
$CuSO_4$ copper sulfate

The effect of polysaccharides extracted from yeast extract on plantlets of *D. lanata* was studied (Ghanem et al. (2010). They revealed that the highest total cardiac glycosides in plantlets tissues (1.427 mg/g D.W.) was observed with 0.1% v/v of polysaccharides and the other treatments had no effect on cardiac glycoside formation. In another study, using temporary immersion systems with elicitors, ChitoPlant and SilioPlant, Perez-Alonso et al. (2012) found that morphological

Table 3.11 Response of in vitro cultures of *Digitalis* to yeast extract and chitin

Species	Culture type	Biotic elicitor	Response	Reference
D. lanata	In vitro plantlets	Yeast extract (0.1, 0.2, 0.4, 0.6 v/v).	Yeast extract at 0.1% gave the highest cardiac glycosides	Ghanem et al. (2010)
D. lanata	In vitro shoots cultivated in temporary immersion systems	Chitoplant: ChP (0.001; 0.01; 0.1 g/l) Silioplant: SiP (0.01; 0.1; 1.0 g/l)	ChitoPlant (0.1 g/l) was found to impact significantly on fresh and dry weight of the shoots The highest accumulation of lanatoside C was achieved with ChitoPlant (0.1 g/l), and SilioPlant (0.01 g/l)	Perez-Alonso et al. (2012)
D. purpurea	Shoot cultures	Yeast extract and commercially available chitin [poly(N-acetyl-1,4-β-D-glucopyranosamine)] at concentrations (0, 100, 200, 300, 400, and 500 mg/l) separately in medium.	Commercial chitin powder and yeast extract could not show any effect on digitoxin accumulation	Patil et al. (2013)
D. purpurea (var. Roter Berggold)	Shoot cultures	ChitoPlant (0,001; 0,01; 0,1 g/l) SilioPlant (0,01; 0,1; 1,0 g/l)	ChitoPlant and SilioPlant (0,01 g/l) increased in 3,6-fold and 6,9-fold digoxin and digitoxin content, respectively	Pérez-Alonso et al. (2014a)

ChitoPlant: strengthens the plants and induces cell wall lignifications as well as production of defense-related enzymes

response of the shoots of *D. lanata* was influenced by elicitors. Regarding biomass production, ChitoPlant (0.1 g/l) was found to impact significantly on fresh and dry weight of the shoots. A higher content of lanatoside C compared to digoxin in all treatments was observed. The highest accumulation of lanatoside C was achieved with ChitoPlant (0.1 g/l), which resulted in 316 μg/g DW and with SilioPlant (0.01 g/l, 310 μg/g DW), which accounted for a 2.2-fold increase in lanatoside C content compared to non-elicited shoot cultures. Additionally, elicitation of shoots resulted in an oxidative stress characterized by hydrogen peroxide and malondialdehyde accumulation. Also, Patil et al. (2013) reported that commercially available yeast extract and chitin powder were inhibitory for the growth in shoot cultures of *D. purpurea*. Commercial chitin powder and yeast extract could not show any effect on digitoxin accumulation. Pérez-Alonso et al. (2014b) showed that ChitoPlant induced a decrease in shoots length, on shoots of *D. purpurea* cultured on semisolid media, but had no effect on the rest of morphological parameters evaluated. Also, ChitoPlant increased cardenolide content. SilioPlant (0.01 g/l) did not affect biomass production but induced the highest net yields per culture flask, 4.72 μg of digoxin and 88.27 μg of digitoxin.

Table 3.12 Response of in vitro cultures of *Digitalis* to fungal elicitors

Species	Culture type	Biotic elicitor	Response	Reference
D. purpurea	Shoot cultures	*Aspergillus niger* (NCIM 545), *Helminthosporium* sp. (NCIM 1280), *Alternaria* sp. (NCIM 1079) In concentrations of (0, 100, 200, 300, 400, and 500 mg/l)	Supplementation of *Helminthosporium* sp. (500 mg/l) improved the digitoxin content by 4.5-fold and digoxin by 3-fold Supplementation of elicitor derived from *Aspergillus niger* could not show any effect on digitoxin accumulation	Patil et al. (2013)

4.2.2 Fungal Elicitors

The use of pathogenic and non-pathogenic fungal preparations as elicitors has become one of the most effective strategies to induce secondary metabolites in most medicinal plants cultured in vitro. Response of in vitro cultures of *Digitalis* to fungal elicitors is presented in Table 3.12.

A study on the effect of fungal elicitor on production of cardenolides from shoot cultures of *D. purpurea* by Patil et al. (2013) was reported. They showed that inclusion of elicitor derived from *Helminthosporium* sp. (500 mg/l) was highly effective in stimulating both digitoxin and digoxin as compared to the control and other elicitors. However, fungal mycelia was inhibitory for the growth of shoots.

4.3 *Precursor Feeding and Cardenolide Production from In Vitro Cultures of* Digitalis *ssp.*

The accumulation of secondary metabolites from callus, suspension cells, and shoot cultures can be enhanced by addition of precursors into the medium. Exogenous supply of a biosynthetic precursor to culture medium may also increase the yield of the desired product. Response of in vitro cultures of *Digitalis* to precursor feeding is presented in Table 3.13.

The effect of methanol, ethanol, propylene glycol, dimethyl sulfoxide and acetone (as the precursor solvents), progesterone, and cholesterol added to the medium on growth and digitoxin content of *D. purpurea* shoot-forming cultures was examined (Hagimori et al. 1983). The results showed that growth was severely repressed by ethanol, promoted slightly by both propylene glycol and dimethyl sulfoxide, and not affected by methanol and acetone. On the other hand, digitoxin production was decreased by all the solvents, most severely by ethanol followed by dimethyl sulfoxide, propylene glycol, methanol, and acetone in that order. Progesterone at 10 mg/100 ml medium or more did repress growth and at 0.1 mg improved the

Table 3.13 Response of in vitro cultures of *Digitalis* to precursor feeding

Species	Culture type	Precursor feeding	Response	Reference
D. purpurea	Cells of shoot-forming cultures	Organic solvents (methanol, ethanol, propylene glycol, dimethyl sulfoxide, and acetone)	Digitoxin production was decreased by all the solvents. Methanol and acetone were relatively the least inhibitive at 2%	Hagimori et al. (1983)
		Progesterone (0.1 to 50 mg/100 ml)	Progesterone at 0.1 mg/100 ml medium improved the digitoxin content per flask to 180% of that of the control cultures	
		Cholesterol (0.1 to 50 mg/100 ml)	No promotive effect	
D. purpurea	Shoot cultures	Progesterone, cholesterol (0, 100, 200, 300, 400, and 500 mg/l) and squalene (0, 1, 2, 3, 4, and 5 mM)	Progesterone (200 to 300 mg/l), in the culture medium, resulted in enhancement accumulation of digitoxin and digoxin by 9.1- and 11.9-fold, respectively	Patil et al. (2013)

digitoxin content per flask to 180% of that of the control cultures, whereas digitoxin formation was repressed by cholesterol at 0.1 mg or more. Patil et al. (2013) found that feeding shoot cultures of *D. purpurea* with precursors associated with the biosynthesis of cardiotonic glycosides such as progesterone, cholesterol, and squalene in the medium resulted in significant decline in growth. The content of digitoxin and digoxin varied significantly with the type and concentration of precursors. The best response for the content of digitoxin (492.6 μg/g DW) and digoxin (214.2 μg/g DW) was on medium fortified with 200 and 300 mg/l of progesterone, respectively. However, when compared to progesterone, cholesterol and squalene were less effective.

5 Accumulation of Cardenolides in In Vitro Culture of *Digitalis* spp.

In this part, the accumulation of cardenolides compounds in different cell suspension cultures, in vitro regenerated plants, and in vivo plants without any stress factors were analysed and compared.

Lui and Staba (1981) studied the effect of culture age and growth regulators on cardenolide content of leaf and root organ cultures of *D. lanata* as well as intact plant leaves. Extracts from leaf cultures, root cultures, and plant roots contained two unidentified genins upon GLC analysis. These unknown genins were not found in extracts of plant leaves.

Hagimori et al. (1982) showed that a considerable amount of digitoxin was found in green shoot-forming cultures of *D. purpurea* than other cultures. Also, cultures of *D. obscura* were established from axillary buds of mature plants or leaves of seedlings under aseptic conditions (Vela et al. 1991). They reported that several cardenolides of series A and C were identified in the regenerants; no significant differences were found in the cardenolide patterns. Regenerated plants from embryogenic leaf cultures grown on MS with 2.7 μM NAA accumulated the highest digoxin (5.5 μg/g dw), while the higher digoxin accumulation (26.0 μg/g dw) was found in plants obtained from axillary buds and grown on MS medium with 0.5 μM BA.

Seitz and Gartner (1994) found that photomixotrophic shoot cultures of *D. purpurea* accumulate cardiac glycosides in substantial concentrations. In extracts of shoot cultures, five enzyme activities are present which are related to the sterol or provide precursors for cardenolide biosynthesis. Three aglycones are accumulated, namely, digitoxigenin, gitoxigenin, and gitaloxigenin. Their time courses of accumulation are very similar. Maximum accumulation is reached after 24 days and then it declines rapidly. The major aglycone is digitoxigenin. Pradel et al. (1997) showed that no cardenolides were detected in hairy roots of *D. lanata* and untransformed roots grown in vitro as well as shoots and plants regenerated from hairy roots. However, anthraquinones and flavonoids were shown to occur. It was concluded that roots of untransformed and transformed plants contained cardenolides that probably were formed in the shoots and transported into the roots.

The development of organ culture based on temporary immersion system can be a reliable method for the steady production of biomass for cardiotonic glycoside production. In this respect, Pérez-Alonso et al. (2009) reported that biomass accumulation of *D. purpurea* shoots grown in temporary immersion system (TIS) was influenced by immersion frequency. The maximum biomass accumulation (106.2 g FW, 11.10 cm shoot length) was obtained with immersions every 4 h (six immersions per day). Digoxin concentrations varied depending on the frequencies tested. In contrast, the digitoxin content showed no dependency on the immersion frequency. Net production of digoxin and digitoxin per TIS was found to be higher with immersions every 4 h. The best net productions of digitoxin and digoxin per TIS were 167.6 and 119.9 μg, respectively. Ghanem et al. (2010) revealed that the high level of digitoxin was found in the in vitro plantlets of *D. lanata* (44.45 μg/g F.W.) followed by the in vivo plants (29.09 μg/g F.W.), while in vivo plant (2 months old) gave the highest value (1.73 μg/g F.W.) of digoxin. Also, Gurel et al. (2011) reported that higher amounts of digoxin accumulation were obtained when shoots of *D. davisiana* were regenerated on LS or Gamborg's B5 medium containing 0.5 mg/l TDZ and 0.25 mg/l IAA, producing 12.59 and 11.93 mg/kg (dw) digoxin, respectively. For natural populations, seasonal variations seemed to affect the production of digoxin in the leaves. The highest amount of digoxin (246.58 mg/kg dw) was in leaf samples collected in July, which coincides with the flowering stage of the plant in the region of collection. Sahin et al. (2013) reported that magnesium deficiency significantly increased cardenolide production in callus cultures of *D. species* tested. However, the increment of cardenolides level promoted by Mg elimination could be possibly associated with putrescine accumulation and significant modifications of enzyme activities related to cardenolide production. In another study, lamina and petiole tissues of regenerated plants of *D. lamarckii* were compared for their cardenolide contents (Yücesan et al. 2014). Lamina extracts showed nearly three times higher cardenolide accumulation than petiole extracts. Of the cardenolides analyzed, neoodorobioside G and glucogitoroside were abundant in lamina extracts (170.3 and 143.9 mg/kg dry weight, respectively).

However, plants regenerated from either callus (R plants) or shoot cultures (S plants) of *D. mariana* were transferred to the greenhouse and propagated for 3, 4, or 10 months (Kreis et al. 2015). Regenerated plants were analyzed for cardenolide content and accumulation pattern. The results showed that the highest cardenolide content, about 2.0 mg/100 g DW in terms of digitoxin equivalents (Dteq), was found in S10 plants (10 months in the greenhouse), whereas plants regenerated from 24-month-old callus (10 months in the greenhouse) had the lowest cardenolide content of about 1.2 mg Dteq/100 g DW. Mohammed et al. (2015) reported that there was no significant difference in lanatoside contents (A, B, and C) between the leaves of *D. cariensis* plants from natural populations and from in vitro derived plantlets grown under greenhouse conditions. Digoxin and digitoxin were not detected in either source. HPLC analysis also revealed that the lanatoside C levels in both natural and in vitro derived plants (106.4 and 100.0 mg/100 g DW, respectively) were higher than those of lanatoside A (62.7 and 59.2 mg/100 g DW, respectively) and lanatoside B (28.7 and 30.7 mg/100 g DW, respectively). However, based on the results presented in this report, in vitro derived plantlets of *D. cariensis* might be a good source for lanatoside A, B, and C production. In a recent study, an efficient in vitro propagation via direct organogenesis was established by Pérez-Alonso et al. (2018). HPLC analysis revealed that contents of digoxin and digitoxin in plants regenerated via direct organogenesis were 22.6 and 220.7 µg/g dw, respectively, while the contents were 20.3 and 218.7 µg/g dw, respectively, in mother plants.

Table 3.14 Accumulation of cardenolides in in vitro culture of *Digitalis* spp.

Species	Tissues examined (Samples analyzed)	Compounds detected	Response	Reference
D. lanata	Extracts of cell suspension derived from leaf and root and in vivo plant leaf and root extracts	Lanatosides A, B, C. Digoxin, digitoxin, and gitoxin Digoxigenin and gitoxigenin Digitoxigenin	The cardenolide patterns from leaf and root culture were similar to those obtained from the in vivo plant	Lui and Staba (1981)
D. purpurea	Undifferentiated, chlorophyllous cell cultures; undifferentiated white cell cultures; green, shoot-forming cultures and white, shoot-forming cultures	Digitoxin	The green cells contained about 6 times more digitoxin than did the white cells. The dark-grown, white, shoot-forming cultures without chloroplasts accumulated about one-third as much digitoxin as did the light-grown, green, shoot-forming cultures	Hagimori et al. (1982)
D. obscura	Plants regenerated through axillary bud culture and leaf cultures	Digoxigenin, deacetyl anatoside C, lanatoside C, and digoxin (series C) and digitoxigenin, deacetyl lanatoside A, and digitoxin (series A)	Digoxigenin derivatives were found in all clonally propagated plants, but the amount of these glycosides was much higher in those obtained from axillary buds	Vela et al. (1991)
D. purpurea	Extracts of axenic, mixotrophic shoot culture	Enzyme extracted: Progesterone 5β-reductase Progesterone 5α-reductase 3β-hydroxysteroid-5β-oxidoreductase Three aglycones are accumulated, namely, digitoxigenin, gitoxigenin, and gitaloxigenin	The progesterone 5α-reductase had an activity maximum (7 days after cultivation), whereas progesterone 5β-reductase activity was highest on day 11 The maximum cardenolide accumulation was after 24 days	Seitz and Gartner (1994)
D. lanata	Hairy roots from infected leaves with *A. rhizogenesis*	Anthraquinones and flavonoids	No cardenolides were detected in hairy roots and untransformed roots grown in vitro. Those compounds were higher in hairy roots than in untransformed roots	Pradel et al. (1997)

(continued)

Table 3.14 (continued)

Species	Tissues examined (Samples analyzed)	Compounds detected	Response	Reference
	Shoot and roots of regenerated plants produced from hairy roots	Digitalinum verum, glucoverodoxin, glucogitoroside, deacetyllanatoside C, neoglucodigifucoside, neoodorobioside G, odorobioside G, lanatoside C, glucoevatromonoside, a-acetyldigoxin, Il-acetyldigoxin, lanatoside A, digitoxin, α-/β-acetyldigitoxin	Shoots and plants regenerated from hairy roots showed similar contents of cardenolides as untransformed shoots and plants	
D. purpurea (cv. Berggold)	Shoots grown in 1.000 ml TIS during 28 days Four immersion frequencies (once every 2, 4, 6, and 12 h)	Digitoxin and digoxin	HPLC analysis revealed the presence of digoxin and digitoxin for all immersion frequencies. No lanatoside C was detected in the biomass cultured in TIS. The best content of digitoxin and digoxin per TIS were 32.5 and 20.6 μg/g dw, after immersions every 2 and 4 h, respectively	Pérez-Alonso et al. (2009)
D. lanata	In vitro produced plantlets and in vivo plants	Digitoxin and digoxin	The digitoxin content reached the maximum value in in vitro plantlet compared with in vivo and in adapted plant, but digoxin content was the highest in in vivo plant	Ghanem et al. (2010)
D. davisiana Heywood	In vitro regenerated plantlets and germinated seedlings	Lanatoside C and digoxin	Digoxin in regenerated plant 12.59 mg/kg dw and in seedlings 14.13 mg/kg dw. Traces of lanatoside C were found in seedlings	Gurel et al. (2011)
	Leaves of plants from natural populations		Higher amounts at flowering stage of the plant (digoxin was 246.58 mg/kg dw)	

D. davisiana Heywood, *D. lamarckii* Ivanina, *D. trojana* Ivanina, and *D. cariensis* Boiss	Callus cultures	Digoxigenin, gitoxigenin, lanatoside C, digoxin, and digitoxin	Higher amounts of five cardenolides, and total cardenolides were obtained when callus were incubated on MS medium lacking both Ca and Mg	Sahin et al. (2013)
D. lamarckii Ivanina	Leaf and petiole of the 4-month-old regenerated plants	Glucogitoroside, strospeside, neo-digitalinum verum, neo-glucodigifucoside, neo-odorobioside G, and glucoevatromonoside	Cardenolide accumulation was higher (nearly threefold) in lamina compared to petiole tissues	Yücesan et al. (2014)
D. mariana Boiss	Leaves of greenhouse-grown in vitro produced plants	Purpurea glycoside A, glucoevatromonoside, lanatoside A, digitoxin, evatromonoside, and α-acetyldigitoxin	The highest content of digitoxin was found in 10-month-old-regenerated plants grown in the greenhouse	Kreis et al. (2015)
D. cariensis	Basal leaves from natural population and in vitro derived regenerants grown under greenhouse conditions for 12 months	Lanatoside (A, B, and C)	No significant difference between extracts of leaves obtained from natural populations and those from in vitro derived regenerants. Digoxin and digitoxin were not detected in either source	Mohammed et al. (2015)
D. purpurea	Plants regenerated via direct organogenesis and mother plants	Digoxin and digitoxin	HPLC analysis revealed that the plants regenerated via direct organogenesis had an appropriate amount of digoxin and digitoxin not significantly different from that of mother plants	Pérez-Alonso et al. (2018)

6 Conclusion

The most prominent compounds formed in the genus *Digitalis* are the cardenolides which are used to treat congestive heart failure. The present chapter includes an up-to-date published work on different digitalis species using biotechnology approaches; micropropagation; direct, indirect regeneration; cryopreservation; and transformation. Effect of abiotic, biotic, and precursor feeding on cardenolide accumulation of several *Digitalis* species were extensively studied.

An efficient plant tissue culture protocol was established with all *Digitalis* species. In most studies, the different explants from *Digitalis* seedling have the potential to regenerate plants through direct or indirect organogenesis or embryogenesis and showed morphogenetic capacity for regeneration, but this capacity depended on the type of explant, culture medium, type and concentration of growth regulators, and cultural conditions. In some reports, it is recommended that somatic embryos are important alternatives to achieve in vitro production of cardenolides. All these regeneration methods have been successfully applied to *Digitalis* species and have been reviewed in depth. Methods for long-term conservation by freezing *Digitalis* cells or tissues in liquid nitrogen were successfully developed. Studies focused on using cell suspension and shoot tips from elite genotypes and successfully cryopreserved using cryogenic storage and encapsulation-dehydration techniques. For the long-term storage of organized structures, the cryopreservation of shoot tips is the method of choice. In the case of suspension cultures, a mixture containing DMSO, glycerol, and sucrose seems to be widely applicable and has also been used with *Digitalis* cultures. RAPD analyses demonstrated that cryopreservation is an efficient method to maintain genetic fidelity.

From the previous research work on tissue culture of *Digitalis* ssp., it is concluded that cell cultures without organ differentiation are unable to produce considerable amount of cardenolides, while shoot or root cultures, in vitro regenerated plants, and even hairy root cultures accumulated sufficient amount of cardenolides than undifferentiated cultures (see Table 3.14). Cardenolide production in *Digitalis* cultures or in vitro plants has proved to be affected by using different abiotic, biotic, and precursor feeding. It was reported that different abiotic elicitors (medium composition, growth regulator levels, photoperiod regime, and chemicals) affect positively on the accumulation of cardenolides in different digitalis cultures. Production of cardenolides (lanatoside C and digoxin) in the materials grown in vitro seemed to correlate with several parameters, such as nutritional and hormonal compositions of the culture medium as well as the duration of culture on the initial regeneration and/or final growth medium. It was found that the elimination of calcium from cell suspension cultures reduced growth and viability of cultures but promoted digoxin formation. An increase of the $MnSO_4$ concentration or the addition of LiCl in the culture medium resulted in higher digoxin content. It was also found that cardenolide accumulation in cell suspension cultures was increased by the ethylene biosynthesis inhibitor aminoethoxyvinylglycine and decreased by the ethylene-releasing substance ethephon. With regard to growth regulators, it was concluded that about twofold increase in digitoxin and 1.8-fold increase in digoxin content was observed in the medium fortified with 5 μM IAA. While the increase in digitoxin was 1.7-fold digoxin was augmented by 1.55-fold on medium containing 5 μM NAA. Methyl jasmonate resulted in decreased biomass production but significantly increased digoxin and digitoxin content. Also, some reports stated that the chloroplasts are not essential for the synthesis of digitoxin in *Digitalis* cells.

With regard to effect of light, it was found that increase in digitoxin content of the callus occurs in the longest period of light exposure. This could be interpreted as light inducing the formation of proplastids, which contain the cardenolide biosynthesis system.

Concerning biotic elicitation, yeast extract and chitin as a biotic elicitors increased the content of digitoxin and digoxin in *Digitalis* shoot cultures by 3–7-fold. Regarding chitoplant, 0.1 g/l was found to impact significantly on biomass production of the *Digitalis* shoots. A higher content of lanatoside C compared to digoxin was observed. Concerning effect of fugal elicitation, it was showed that supplementation of *Helminthosporium* sp. (500 mg/l) improved the digitoxin content by 4.5-fold and digoxin by 3-fold as compared to the control treatment. However, fungal mycelia was inhibitory for the growth of *Digitalis* shoots. Exogenous supply of a biosynthetic precursor to culture medium may also increase the yield of the desired product. In some reports, the content of digitoxin and digoxin of shoot cultures was increased 9–12-fold by adding 0.1 mg/100 ml progesterone in the culture medium. In general, more detailed knowledge of the enzymes and genes involved in cardenolide biosynthesis is essential for studying the regulation and engineering of the cardenolide pathway in the future.

7 Future Aspects

At present, cultivation of foxglove still remains the only source of cardenolides. They are still extracted from plants grown in the field, though attempts have been made in the past to produce cardenolides by plant biotechnology techniques. Promising results were successfully reported by several researchers on the establishment of protocols for cell and transgenic cultures of *Digitalis* species. Molecular and enzymology studies on different cell and tissue cultures and biotransformation reactions are still important research areas for production of cardenolides. Besides these, metabolic engineering for *Digitalis* species with genes involved in the cardenolide biosynthesis, for the production of digoxin, is extensively required.

For thousands of years, plants were the major source of natural products for human medicines. The difficulty in resynthesizing these products, however, often turned pharmaceutical industries away from this rich source for human medicine. More recently, progress on transformation through genetic manipulation of biosynthetic units in microorganisms has opened the possibility of in-depth exploration of the large chemical space of natural products derivatives (Trosset and Carbonell 2015). Concerning synthetic biology and cardenolides, recent work of Munkert et al. (2016) are about to engineer yeast and *E. coli* for producing different cardenolides. They combine enzyme discovery, enzyme engineering, as well as pathway optimization to realize this work. With the recent advanced genome editing, molecular biology, and protein engineering tools, synthetic biology technology has become a powerful tool for creating biological devices and increasing the flux of compounds of interest.

References

Aawad ZJ, Al-Khateeb EH (2006) Using tissue culture technique for the production of cardiac glycosides from roots of *Digitalis purpurea* L plantlets (Var. Excelsior Mixed). Iraqi J Pharm 15(1):92–98

Arrillaga I, Brisa MC, Segura J (1987) Somatic embryogenesis from hypocotyl callus cultures of *Digitalis obscura L.* Plant Cell Rep 6:223–226

Berggren B, Ohlsson AB (1991) Ultrastructure of *Digitalis lanata* tissue cultures. Effect of gibberellic acid and SAN 9789. Protoplasma 161:58–69

Berglund T, Ohlsson AB (1992) Effects of ethylene and aminoethoxyvinylglycine on cardenolide accumulation in tissue cultures of *Digitalis lanata*. J Plant Physiol 140:395–398

Bosila HA, Mohamed S, Gamal SE, Bekhit M (2003) Factors affecting callus production and glycosidal content of leaf tissue culture of *Digitalis lanata* Ehrh. Acta Hortic 597:289–301

Bota C, Deliu C (2011) The effect of copper sulphate on the production of flavonoids in *Digitalis lanata* cell cultures. Farmacia 59(1):113–118

Brauchler C, Meimberg H, Heubl G (2004) Molecular phylogeny of the genera *Digitalis* L. and Isoplexis (Lindley) Loudon (Veronicaceae) based on ITS and *trn*L-F sequences. Plant Syst Evol 248:111–128

Cacho M, Moran M, Herrera MM, Fernandez-Tarrago J, Corchete MP (1991) Morphogenesis in leaf, hypocotyl and root explants of *Digitalis thapsi L.* cultured *in vitro*. Plant Cell Tissue Organ Cult 25:117–123

Cacho M, Moran M, Fernandez-Tarrago J, Corchete P (1995) Calcium restriction induces cardenolide accumulation in cell suspension cultures of *Digitalis thapsi L.* Plant Cell Rep 14:786–789

Cacho M, Moran M, Corchete P, Fernandez-Tarrago J (1999) Effect of calcium restriction on cardenolide accumulation in two cell lines of *Digitalis thapsi* grown under different light regimes. Acta Physiol Plant 21(4):335–340

Chavan SP, Nitnaware KM, Lokhande VH, Nikam TD (2011) Influence of growth regulators and elicitors on cell growth and α-tocopherol and pigment productions in cell cultures of *Carthamus tinctorius* L. Appl Microbiol Biotechnol 89:1701–1707

Corchete MP, Jimenez MA, Moran M, Cacho M, Fernandez-Tarrago J (1991) Effect of calcium, manganese and lithium on growth and cardenolide content in cell suspension cultures of *Digitalis thapsi L.* Plant Cell Rep 10:394–396

Corduan G, Spix C (1975) Haploid callus and regeneration of plants from anthers of *Digitalis purpurea L.* Planta (Berl) 124:1–11

Corduk NE, Aki C (2010) Direct shoot organogenesis of *Digitalis trojana* Ivan., an endemic medicinal herb of Turkey. Afr J Biotechnol 9(11):1587–1591

Diettrich B, Popov AS, Pfeiffer B, Neumann D, Butenko R, Luckner M (1982) Cryopreservation of *Digitalis lanata* cell cultures. Planta Med 46(10):82–87

Eisenbeiss M, Kreis W, Reinhard E (1999) Cardenolide biosynthesis in light- and dark-grown *Digitalis lanata* shoot cultures. Plant Physiol Biochem 37(1):13–23

Elbaz HA, Stueckle TA, Wang HL, O'Doherty GA, Lowry DT, Sargent LM, Wang L, Dinu CZ, Rojanasakul Y (2012) Digitoxin and a synthetic monosaccharide analog inhibit cell viability in lung cancer cells. Toxicol Appl Pharmacol 258:51–60

Fatima Z, Mujib A, Fatima S, Arshi A, Umar S (2009) Callus induction, biomass growth, and plant regeneration in *Digitalis lanata* Ehrh.: influence of plant growth regulators and carbohydrates. Turk J Bot 33:393–405

Gartner DE, Seitz HU (1993) Enzyme activities in cardenolide-accumulating, mixotrophic shoot cultures of *Digitalis purpurea* L. J Plant Physiol 141:269–275

Ghanem SA, Aboul-Enein AM, El-Sawy A, Rady MR, Ibrahem MM (2010) *In vitro* propagation and cardiac glycosides content of *Digitalis lanata*. Int J Acad Res 2(6):349–356

Goldner EM, Seitz U, Reinhard E (1991) Cryopreservation of *Digitalis lanata* Ehrh. cell cultures: preculture and freeze tolerance. Plant Cell Tissue Organ Cult 24:19–24

Gurel E, Yucesan B, Aglic E, Gurel S, Verma SK, Sokmen M, Sokmen A (2011) Regeneration and cardiotonic glycoside production in *Digitalis davisiana* Heywood (Alanya Foxglove). Plant Cell Tissue Organ Cult 104:217–225

Hagimori M, Matsumoto T, Obi Y (1982) Studies on the production of *Digitalis* cardenolides by plant tissue culture. II. Effect of light and plant growth substances on digitoxin formation by undifferentiated cells and shoot-forming cultures of *Digitalis purpurea* L. grown in liquid media. Plant Physiol 69:653–656

Hagimori M, Matsumoto T, Obi Y (1983) Effects of mineral salts, initial pH and precursors on digitoxin formation by shoot-forming cultures of *Digitalis purpurea* L. grown in liquid media. Agric Biol Chem 47(3):565–571

Herrera MT, Cacho M, Corchete MP, Fernandez-Tarrago J (1990) One step shoot tip multiplication and rooting of *Digitalis thapsi L.* Plant Cell Tissue Organ Cult 22:179–182

Koga M, Hirashima K, Nakahara T (2000) The transformation system in foxglove (Digitalis purpurea L.) using *Agrobacterium rhizogenes* and traits of the regenerants. Plant Biotechnol 17(2):99–104

Kreis W (2017) The foxgloves (digitalis) revisited. Planta Med 83:962–976

Kreis W, Hensel KA, Stuhlemmer U (1998) Cardenolide biosynthesis in foxglove. Planta Med 64:491–499

Kreis W, Haug B, Yücesan B (2015) Somaclonal variation of cardenolide content in Heywood's foxglove, a source for the antiviral cardenolide glucoevatromonoside, regenerated from permanent shoot culture and callus. In Vitro Cell Dev Biol Plant 51:35–41

Lapena L, Perez-Bermudez P, Segura J (1992) Factors affecting shoot proliferation and vitrification in *Digitalis obscura* cultures. In Vitro Cell Dev Biol 28:121–124

Lehmann U, Moldenhauer D, Thomar S, Diettrich B, Luckner M (1995) Regeneration of plants from *Digitalis lanata* cells transformed with *Agrobacterium tumefaciens* carrying bacterial genes encoding neomycin phosphotransferase II and β-glucuronidase. J Plant Physiol 147:53–57

Li Y, Gao Z, Piao C, Lu K, Wang Z, Cui ML (2014) A stable and efficient *Agrobacterium tumefaciens*-mediated genetic transformation of the medicinal plant *Digitalis purpurea* L. Appl Biochem Biotechnol 172(4):1807–1817

Linsmaier E, Skoog F (1965) Organic growth factor requirements of tobacco tissue cultures. Physiol Plant 18:100–127

Luckner M, Wichtl M (2000) Digitalis – Geschichte, Biologie, Chemie, Physiologie, Molekularbiologie, Pharmakologie, medizinische Anwendung. Wissenschaftliche Verlagsgesellschaft, Stuttgart

Lui JHC, Staba EJ (1981) Effects of age and growth regulators on serially propagated *Digitalis lanata* leaf and root cultures. Planta Med 41:90–95

Mohammed A, Yücesan B, Demir-Ordu O, Cihangir C, Eker I, Kreis W, Gürel E (2015) *In vitro* regeneration and cardenolide determination of an endemic foxglove, *Digitalis cariensis* (Aegean Foxglove). In Vitro Cell Dev Biol Plant 51:438–444

Munkert J, Geiger D, Meitinger N, Rieck C, Petersen J, Kreis W (2016) Production of natural and semisynthetic cardenolides–a synthetic biology approach. Abstracts/New Biotechnol 33:S1–S213. https://doi.org/10.1016/j.nbt.2016.06.1358

Murashige T, Skoog F (1962) A revised medium for rapid growth and bioassays with tobacco tissue cultures. Physiol Plant 15:473–497

Nitseh JP, Nitsch C (1969) Haploid plants from pollen grains. Science 163:85–87

Ohlsson AB (1990) Effects of abscisic acid on cardenolide accumulation and growth in *Digitalis lanata* tissue cultures. J Plant Physiol 136:510–512

Ohlsson AB, Berglund T (1989) Effects of high $MnSO_4$ levels on cardenolide accumulation by *Digitalis lanata* tissue cultures in light and darkness. J Plant Physiol 135:505–507

Ohlsson AB, Bjork L (1988) Effect of gibberellic acid on cardenolide accumulation by *Digitalis lanata* tissue culture grown in light and darkness. J Plant Physiol 133:535–538

Ohlsson AB, Bjork L, Gatenbeck S (1983) Effect of light on cardenolide production by *Digitalis lanata* tissue cultures. Phytochemistry 22(11):2447–2450

Pádua RM, Meitinger N, Dias de Souza JF, Waibel R, Gmeiner P, Braga FC, Kreis W (2012) Biotransformation of 21-O-acetyl-deoxycorticosterone by cell suspension cultures of *Digitalis lanata* (strain W.1.4). Steroids 77:1373–1380

Patil JG, Ahire ML, Nitnaware KM, Panda S, Bhatt VP, Kishor PBK, Nikam TD (2013) *In vitro* propagation and production of cardiotonic glycosides in shoot cultures of *Digitalis purpurea* L. by elicitation and precursor feeding. Appl Microbiol Biotechnol 97:2379–2393

Pérez-Alonso N, Wilken D, Gerth A, Jahn A, Nitzsche HM, Kerns G, Capote-Perez A, Jiménez E (2009) Cardiotonic glycosides from biomass of *Digitalis purpurea* L. cultured in temporary immersion systems. Plant Cell Tissue Organ Cult 99:151–156

Perez-Alonso N, Capote A, Gerth A, Jimenez E (2012) Increased cardenolides production by elicitation of *Digitalis lanata* shoots cultured in temporary immersion systems. Plant Cell Tissue Organ Cult 110:153–162

Pérez-Alonso N, Chong-Perez B, Capote A, Perez A, Izquierdo Y, Angenon G, Jimenez E (2014a) *Agrobacterium tumefaciens*-mediated genetic transformation of *Digitalis purpurea* L. Plant Biotechnol Rep 8:387–397

Pérez-Alonso N, Labrada FA, Capote A, Pérez A, Sosa R, Mollineda A, Gonzalez EJ (2014b) Stimulation of cardenolides production in *Digitalis purpurea* L. shoot cultures by elicitors addition. Rev Colomb Biotecnol XVI(1):51–61

Pérez-Alonso N, Martín R, Capote A, Pérez A, Hernández-Díaz EK, Rojas L, Jiménez E, Quiala E, Angenon G, Garcia-Gonzales R, Chong-Pérez B (2018) Efficient direct shoot organogenesis, genetic stability and secondary metabolite production of micropropagated *Digitalis purpurea* L. Ind Crop Prod 116:259–266

Perez-Bermudez P, Brisa MC, Cornejo MJ, Segura J (1984) *In vitro* morphogenesis from excised leaf explants of *Digitalis obscura* L. Plant Cell Rep 3:8–9

Perez-Bermudez P, Garcia AAM, Tunon I, Gavidia I (2010) *Digitalis purpurea* P5βR2, encoding steroid 5β-reductase, is a novel defense-related gene involved in cardenolide biosynthesis. New Phytol 185:687–700

Pradel H, Dumke-Lehmann U, Diettrich B, Luckner M (1997) Hairy root cultures of *Digitalis lanata.* Secondary metabolism and plant regeneration. J Plant Physiol 151:209–215

Roden DM (2001) Drugs effecting renal and cardiovascular system. In: Gilman GA, Hardman JG, Limbird LE (eds) The pharmacological basis of therapeutics. McGraw Hill, New York, p 920

Sahin G, Verma SK, Gurel E (2013) Calcium and magnesium elimination enhances accumulation of cardenolides in callus cultures of endemic *Digitalis* species of Turkey. Plant Physiol Biochem 73:139–143

Saito K, Yamazaki M, Shimomura K, Yoshimatsu K, Murakoshi I (1990) Genetic transformation of foxglove *(Digitalis purpurea)* by chimeric foreign genes and production of cardioactive glycosides. Plant Cell Rep 9:121–124

Sales E, Nebauer SG, Arrillaga I, Segura J (2001) Cryopreservation of *Digitalis obscura* selected genotypes by encapsulation-dehydration. Planta Med 67:833–838

Sales E, Segura J, Arrillaga I (2003) *Agrobacterium tumefaciens*-mediated genetic transformation of the cardenolide-producing plant *Digitalis minor* L. Planta Med 69:143–147

Seitz HU, Gartner DE (1994) Enzymes in cardenolide-accumulating shoot cultures of *Digitalis purpurea* L. Plant Cell Tissue Organ Cult 38:337–344

Seitz U, Alfermann AW, Reinhard E (1983) Stability of biotransformation capacity in *Digitalis lanata* cell cultures after cryogenic storage. Plant Cell Rep 2:273–276

Trosset JY, Carbonell P (2015) Synthetic biology for pharmaceutical drug discovery. Drug Des Devel Ther 9:6285–6302

Vela S, Gavidia I, Perez-Bermijdez P, Segura J (1991) Micropropagation of juvenile and adult *Digitalis obscura* and cardenolide content of clonally propagated plants. In Vitro Cell Dev Biol 27:143–146

Verma SK, Yucesan BB, Şahin G, Gurel S, Gurel E (2011a) Direct shoot regeneration from leaf explants of *Digitalis lamarckii*, an endemic medicinal species. Turk J Bot 35:689–695

Verma SK, Yucesan BB, Gurel S, Gurel E (2011b) Indirect somatic embryogenesis and shoot organogenesis from cotyledonary leaf segments of *Digitalis lamarckii* Ivan., an endemic medicinal species. Turk J Biol 35:743–750

Verma SK, Sahin G, Yucesan B, Eker I, Sahbaz N, Gurel S, Gurel E (2012) Direct somatic embryogenesis from hypocotyl segments of *Digitalis trojana* Ivan and subsequent plant regeneration. Ind Crop Prod 40:76–80

Verma SK, Sahin G, Gurel E (2016) Somatic embryogenesis, pigment accumulation and synthetic seed production in *Digitalis davisiana* Heywood. Indian J Exp Biol 54:245–253

Verma SK, Das AK, Gantait S, Gurel S, Gurel E (2018) Influence of auxin and its polar transport inhibitor on the development of somatic embryos in *Digitalis trojana*. 3 Biotech 8:99

White PR (1943) A handbook of plant tissue culture. Jaques Cattell Press, Tempe

Yücesan B, Müller-Uri F, Kreis W, Gürel E (2014) Cardenolide estimation in callus-mediated regenerants of *Digitalis lamarckii* Ivanina (dwarf foxglove). In Vitro Cell Dev Biol Plant 50:137–142

Yücesan B, Mohammed A, Eker I, Sameeullah M, Demir-Ordu O, Cihangir C, Şahbaz N, Kaya O, Müller-Uri F, Kreis W, Gürel E (2016) *In vitro* propagation and cardenolide profiling of *Digitalis ferruginea* subsp. *schischkinii*, a medicinally important foxglove species with limited distribution in Northern Turkey. In Vitro Cell Dev Biol Plant 52:322–329

Yücesan BB, Eker I, Lazzarini LES, Aslam N, Mohammed A, Pinto JEBP, Kreis W, Gürel E (2018) Shoot-tip cultivation and cardenolide content analysis of natural populations of *Digitalis lanata* Ehrh. *subsp. lanata* (wooly foxglove) in Thrace region. Int J Agric Wildl Sci 4:55–62

Index

M. R. Rady, *Plant Biotechnology and Medicinal Plants*,
https://doi.org/10.1007/978-3-030-22929-0

Kolibri 360 GmbH
Ferdinand-Jühlke-Straße 7
99095 Erfurt, Deutschland
produktsicherheit@kolibri360.de